Hazardous Waste Management

Edited by Rajesh Banu Jeyakumar, Kavitha Sankarapandian and Yukesh Kannah Ravi

Published in London, United Kingdom

Hazardous Waste Management
http://dx.doi.org/10.5772/intechopen.96834
Edited by Rajesh Banu Jeyakumar, Kavitha Sankarapandian and Yukesh Kannah Ravi

Contributors
Mitsuo Yoshida, Charlene Nagawa, Tomonori Ishigaki, Hiroyuki Ishimori, Hiroki Kitamura, Masato Yamada, Yıldırım İsmail Tosun, Olga Muter, Laila Dubova, Oleg Kassien, Jana Cakane, Ina Alsina, Sahabudeen Sheik Mohideen, Prathyusha Cota, Sayantani Saha, Shailvi Tewari, Abhirami Sasikumar, M. Yashwant Saran, Swetha Senthilkumar, Ifeanyi Michael Smarte Anekwe, Donald Tyoker Kukwa, Felicia Omolara Afolabi, Emmanuel Kweinor Tetteh, Maggie Manimagalay Chetty, Innocent Chukwunonso Chukwunonso Ossai, Fauziah Shahul Hamid, Auwalu Hassan, Amara Dar, Arooj Naseer

First published in London, United Kingdom, 2022 by IntechOpen
IntechOpen is the global imprint of INTECHOPEN LIMITED, registered in England and Wales, registration number: 11086078, 5 Princes Gate Court, London, SW7 2QJ, United Kingdom

British Library Cataloguing-in-Publication Data
A catalogue record for this book is available from the British Library

Additional hard and PDF copies can be obtained from orders@intechopen.com

Hazardous Waste Management
Edited by Rajesh Banu Jeyakumar, Kavitha Sankarapandian and Yukesh Kannah Ravi
p. cm.
Print ISBN 978-1-80355-378-8
Online ISBN 978-1-80355-379-5
eBook (PDF) ISBN 978-1-80355-380-1

Meet the editors

Dr. J. Rajesh Banu is an associate professor in the Department of Life Sciences, Central University of Tamil Nadu (CUTN), India. He is also a visiting professor at the Center for Excellence in Environmental Studies, King Abdul Aziz University, Saudi Arabia. He has seventeen years of teaching experience at national (Anna University and CUTN) and international universities (Sungkyunkwan University (SKKU), South Korea). He is a visiting researcher at East China Normal University, China, and a former post-doctoral fellow at SKKU. He is the author of 250 publications in SCI and SCIE journals and a senior editor of e-Prime. He has authored three books, fifty book chapters, and fourteen conference proceedings. He obtained seven major grants from the Government of India and two consultancies from industries. He has guided seventeen Ph.D. students and 100 postgraduate engineering students in the discipline of environmental engineering and science. Dr. Banu is on Stanford University's list of the World's Top 2% Scientists and the recipient of the 2018 Hiyoshi Think of Ecology Award. His research centers on waste to energy, value-added product recovery from waste, biological wastewater treatment, nutrient removal, membrane bioreactors, and microbial fuel cells.

Dr. Kavitha Sankarapandian obtained her Ph.D. in Environmental Biotechnology discipline from Anna University, Chennai, India. She received her bachelor's degree in Microbiology and a master's in Biotechnology from Manonmaniam Sundaranar University, Tirunelveli, India. Her research interests include the generation of energy from biomass, micro and macroalgae biorefinery, lignin valorization, and microbial fuel cells. She has authored seventy-six articles in SCI and SCIE journals, one book, and fourteen book chapters. She is also a reviewer and editorial board member for various international journals. She has received three best poster awards at international conferences. Dr. Kavitha was ranked among the World's Top 2% Scientists by Stanford University in 2020.

Dr. Yukesh Kannah Ravi is a visiting assistant professor in the Department of Environmental and Sustainable Engineering, State University of New York at Albany, USA. He obtained a bachelor's degree in Civil Engineering, a master's degree in Environmental Engineering, and a Ph.D. in Environmental Engineering from Anna University, Chennai, India. His research focuses on the conversion of waste biomass to bioenergy/biofuels, biological wastewater treatment, biomass biorefinery, and value-added product recovery. He has authored and co-authored forty-five peer-reviewed scientific journal articles and twenty book chapters. He is also a reviewer and editorial board member for various international journals. Dr. Yukesh has received three best poster awards at international conferences.

Contents

Preface

Rapid growth in industrialization and expansion of innovative technologies in chemical processing are having negative impacts on the environment and public health. Nowadays, new products and pollutants are increasing daily due to drastic changes in human activities. Toxic, non-biodegradable, radioactive, and flammable substances are collectively known as "hazardous waste." Managing this type of waste is a challenging task for scientists and engineers. Typically, physicochemical and biological treatment techniques are employed to manage hazardous waste, however, current technologies are not capable or effective enough to treat contaminants in compliance with standards. Thus, the integration of different technologies is an emerging area of research and development for the management and treatment of hazardous waste. This book discusses the most important techniques of hazardous waste management practices, bioremediation, and value-added product recovery. It provides information on engineering measures and technical challenges to be considered for successful hazardous waste management. This book is a useful resource for waste management and treatment professionals, chemical engineers, technicians, medical professionals, and environmental regulators as well as students studying hazardous waste management, environmental engineering, and environmental science.

Rajesh Banu Jeyakumar
Department of Life Sciences,
Central University of Tamil Nadu,
Thiruvarur, Tamil Nadu, India

Kavitha Sankarapandian
Former Researcher,
Department of Civil Engineering,
Anna University Regional Campus Tirunelveli,
Tirunelveli, Tamilnadu, India

Yukesh Kannah Ravi
Department of Environmental and Sustainable Engineering,
University at Albany,
The State University of New York,
Albany, New York, United State of America

Section 1

Hazardous Waste

Chapter 1

Acrylamide: A Neurotoxin and a Hazardous Waste

Prathyusha Cota, Sayantani Saha, Shailvi Tewari, Abhirami Sasikumar, M. Yashwant Saran, Swetha Senthilkumar and Sahabudeen Sheik Mohideen

Abstract

Acrylamide is an organic water-soluble compound and a vinyl-substituted primary amide. It is well known for its toxic effects on humans. This chemical may lead to neurodegenerative disorders like Alzheimer's and Parkinson's. It is exposed to humans through diet, occupation, lifestyle and many environmental factors. Acrylamide is used in molecular laboratories and even in various manufacturing and processing industries. Acrylamide is formed in food cooked at high temperatures, and exposure to this chemical may cause damage to the nervous system. In this chapter the toxicity of acrylamide and its role as a hazardous waste are highlighted. The main topics of this study are occurrence, effects and toxicity caused by acrylamide and analysis of acrylamide induced neurotoxicity in rats. Furthermore, mitigation strategies involving acrylamide have been discussed.

Keywords: acrylamide, hazardous waste, neurotoxicity, oxidative stress, therapeutic agents, mitigation

1. Introduction

Acrylamide (ACR), a water-soluble vinyl monomer, is a by-product of foods rich in carbohydrates that are cooked at higher temperatures. It has been shown to evoke genotoxic, carcinogenic, and neurotoxic effects in various kinds of animal species [1]. When exposed to humans through lifestyle, diet, occupation, and various other environmental factors it can cause adverse neurotoxic effects like ataxia, peripheral neuropathy and may result in the pathogenesis of neurodegenerative diseases. ACR could be found as a result of the use of polyacrylamide (PAM) in the environment. It can be seen in ingredients that are eaten by humans daily, including biscuits, breakfast cereals, bread, and crackers [2]. They are also used in cosmetics and toiletries, paper and textile production, production of dyes and organic chemicals, sugar refining, etc. PAM depolymerizes when it is exposed to high temperatures or pH to form ACR causing contamination to the environment [3]. Some other ways of exposure to ACR are through oral, inhalation, and dermal routes. It is formed from the amino acid asparagine during high-temperature cooking like baking and frying. Studies have reported that ACR is obtained from reducing sugars and the amino acid asparagine through the Maillard reaction [2]. Asparagine has been considered to be a major precursor of ACR, and heating foods having high starch content such as potatoes can result in high levels of asparagine eventually resulting in high ACR formation [4]. Rats subjected to specific time and dose-dependent measures of ACR have shown decreased norepinephrine levels and density of noradrenergic axons in different parts of the brain showing morphological evidence. According to the US FDA, a survey of 2015 on ACR values in individual food product samples, 70 ppb of ACR level was found for bread and bakery products, 500 ppb for Nuts and Fruits, and 1030 ppb for French Fries and Other Potato Foods. The no-observed-adverse-effect limit (NOAEL) and lowest observed adverse effect level (LOAEL) for laboratory animals is 0.2–0.5 μg/kg/day and 2 μg/kg/day respectively whereas the mean dietary exposure estimated by the World Health Organisation (WHO) is 0.001 cmg/kg/day [5]. It is, therefore, crucial to identify the cause of and exposure to ACR, its ways of reduction, and the health risks that are involved to establish a safer environment. The European Food Safety Authority (EFSA) has estimated the Benchmark Dose Lower Confidence Limit (BMDL) for ACR. For tumours, experts chose a $BMDL_{10}$ of 0.17 mg/kg bw/day and for other effects, neurological alterations were seen to be closely related with a $BMDL_{10}$ of 0.43 mg/kg bw/day [6]. Currently, many mitigation strategies are being investigated for their therapeutic effects against ACR present in the environment. Due to the harmful effects of ACR, research focuses on human health risks, dietary exposure to ACR, and its limit in foods by modulating processing ACR [3, 4].

Although the exact mechanism of ACR toxicity is still under investigation, many studies have shown that an imbalance in the antioxidant system can be one of the major reasons [7]. However, the US FDA [8] suggests that the level of ACR used in laboratory studies is higher than what humans are exposed through food. Also, the study by [9] states that more investigation is required to establish occupational exposure levels of ACR.

This review chapter aims to throw light on ACR as a neurotoxin and hazardous waste by discussing various aspects like the occurrence of ACR as a hazardous waste, effects of ACR and its by-products, ACR induced neurotoxicity leading to neurodegenerative changes and the potential of different therapeutic strategies to mitigate the toxicity.

2. Occurrence of acrylamide as a hazardous waste

ACR is a monomer and may be found in the environment because of the use of PAM polymers. ACR and its derivatives are used as sewage-flocculating agents and mainly occur in mineral extraction and chemical and food processing industries [10]. PAMs are agents also used in soil conditioning and strengthening in paper manufacturing [11]. ACR contaminates water through the use of PAM polymers in a range of industries such as agricultural, oil drilling, cement, herbicide, paper production, cosmetics, soap, chalk, adhesives, dyes, explosives, printing inks, and latex. All of these PAM applications, particularly flocculants and soil stabilisers, are potential sources of PAM contamination in drinking water supplies. Toxicity testing with some PAM-sensitive aquatic organisms revealed that oil-based PAM was harmful, whereas water-based PAM products were not. The cationic PAM has a lethal concentration (LC50a) of 0.3–10 mg/L, and it adheres to fish gills, obstructing the osmoregulation system. Several studies have found that anionic PAM products are safer to use in environmental water than cationic and neutral PAM products. Rainbow trout, especially larger fish rather than fingerlings, have seen acute alterations in their gills at LC as a result of cationic PAM poisoning. Due to ACR exposure, goldfish developed acute tissue lesions in the pancreas and genotoxic damages in their erythrocytes, disrupting homeostasis and eventually having a carcinogenic effect. It is also noted that ACR is not accumulated in sludges produced by PAM flocculants [10]. At room temperature, ACR is a solid, however it is extremely soluble (2155 g L1 of water) and mobile in water [12]. The major source of ACR in drinking water is the residual monomer of PAM, which is released throughout the treatment process. PAMs can be a source of release to drinking water sources when used as a chemical grouting agent and soil stabiliser in the building of tunnels, sewers, wells, reservoirs, and dams. ACR is also released into water by plastic and dye industries. Because ACR does not participate in soil binding but is extremely soluble and mobile in water, it will travel quickly with seepage, increasing the risk of pollution of surface or groundwater [13]. The concentration of ACR in aquatic and terrestrial ecosystems around ACR or PAM using—industries was found to be 0.3 ppb to 5 ppm [14].

Individuals inhale ACR mostly through smoking [15]. The amount of haemoglobin adduct identified was precisely related to the amount of ACR inhaled from three cigarettes each day [16]. Continuous ACR exposure, particularly by blue-collar workers, has been linked to headaches, muscle weakness, increased sensitivity in their extremities, dyspnea, and in certain cases, balance impairment, paresthesia, discomfort, and truncal ataxia [13]. The average ACR level per cigarette is 679.3 ng, with a range of 455.0–822.5 ng per cigarette. Adult smokers in Poland are predicted to be exposed to 0.17 g/kg b.w. of ACR per day via tobacco smoke [17]. It has been stated that each cigarette contains an average of 1.2 μg of ACR. According to the study, smoking 20 cigarettes per day exposes the body to 0.5 μg/kg b.w. per day [18]. ACR poisoning causes rashes, peeling of the skin and hands, cramping, and sweating, among other symptoms. It's also a skin irritant that causes peeling contact dermatitis on the palms, which can lead to neurologic conditions. In humans, dermal exposure can cause an exfoliative reddish rash [13, 19, 20]. In a study including two grouting workers, it was observed that one has experienced skin peeling after 2 weeks of exposure to high concentration of ACR and systemic neuropathy in the next 6 months whereas the other worker showed cerebellar dysfunctions, including gait ataxia and slurred speech after 1-month exposure [19]. The use of ACR in cosmetics is a risk to the population. Due to ACR toxicity, the initial approved dose of 100 $mg.kg^{-1}$ cosmetic product was reduced to 0.5 $mg.kg^{-1}$ therefore the daily exposure due to cosmetics has lowered to 0.7 $g.kg^{-1}$ b.w. per day.

3. Effects of acrylamide and its by-products

3.1 Reproductive toxicity

ACR can play a direct role in the toxicological effects of sperm morphology, motility, and production, as well as being an indirect cause of reproductive issues as shown by various studies done on ACR exposed male mice [16, 19]. Repeated injections of ACR (20 mg/kg) into male rats for 20 days resulted in dose-dependent reductions in testosterone and prolactin levels [21]. In another study it was observed, after a one-month experiment, ACR exposure at levels of 1.25–24 mg/kg/day in their drinking water lowered fertility rates and litter sizes in mice, while increasing morphological anomalies of sperm and embryo resorption rates. In animal toxicity experiments using ACR, decreased reproductive behaviour, testicular atrophy, aberrant spermatogenesis, and poor sperm quality are some of the symptoms [13]. Peripheral neuropathies caused by ACR, such as decreased hind-limb function, may impede copulatory behaviour, mounting responses, and intromission, eventually impact the sperm deposition in the vagina and uterus, as well as cause hormonal alterations. In terms of hormonal mechanisms of action, ACR decreased serum testosterone and prolactin levels, which could contribute to testicular shrinkage and sperm motility [22].

3.2 Genotoxicity, carcinogenicity and mutagenicity

When the nervous systems of humans and animals are exposed to excessive quantities of ACR, α, β unsaturated carbonyl molecule with strong chemical activity, it can cause cancer and neurotoxicity [23]. ACR and its metabolites have been shown to be both genotoxic and carcinogenic in various studies [24, 25]. When ACR enters the body, it is oxidised and transformed into the genotoxic metabolite glycidamide (GA) [26]. ACR is ingested in the digestive system and transported to the liver at a rate of 4 mol ACR per 1 mol haemoglobin, where it is processed and destroyed by two distinct routes. The carcinogenic action of ACR, which is metabolised to GA in the liver by CYP2E1, has a mutagenic effect in the brain, kidneys, lungs, uterus, and testis in several organisms that includes experimental animals as well [27]. It was discovered that the genotoxic effect on DNA was mostly caused by GA, an ACR metabolite, rather than ACR itself [28]. The conversion of ACR to GA was found to be quite common in rats and mice, and it's mode of action included it's interaction with purine bases in the liver, renal, and pulmonary DNAs of rats and mice, causing genotoxic impacts [24, 25]. ACR also induced gene mutations and chromosomal defects in cultured mouse embryonic fibroblast cells, according to in vitro experiments [29, 30].

In this chapter, we have focused on neurotoxic effects caused by ACR in rats. ACR is predominantly known as a neurotoxin in humans. In this chapter, we are discussing ACR induced neurotoxicity in rat models where extensive studies have been done.

4. Analysis of acrylamide induced neurotoxicity in rat models

4.1 Dose and time dependent response

Studies have been made to understand the relation between the dose response and effect of ACR in rats. ACR has shown remarkable toxic effects with acute doses from the very beginning [31]. The response has been found to vary with the dose of exposure. The different doses of ACR to which the rat model has been exposed to be

No. of days of exposure	Dose of exposure (mg/kg)	Effects	References
24 hours	0.5, 2.5, 12.5	No apoptotic neuronal death, decrease in GSH.	[1]
10 days	38.27 (1/3rd dose of LD50)	Decreased in GSH, SOD, CAT and AChE activity and increase in LPO.	[36]
12 days	40	Weight loss, gait abnormality, Purkinje cell nuclear condensation, DNA damage in rat cerebellum were observed after the exposure period.	[1]

No. of days of exposure	Dose of exposure (mg/kg)	Effects	References
21 days (3 weeks)	25	A decrease in the haematological parameters, brain NT concentrations, AChE activity, antioxidant biomarkers. Elevation in the levels of oxidative stress biomarkers. Astrocytosis was also observed.	[7]
28 days (4 weeks)	5	Increase in weight but no neurotoxicity observed	[33, 35]
	15, 30	Uncoordinated motor movement, nervous function defects, increase in the quantity of abnormal neurons distributed in varied layers of the cerebral cortex and wide distribution of astrocytes in the brain was observed.	
	40	A significant loss in body weight, continuing deficits in motor function, adverse pathological changes in the cortex and hippocampus of rats.	
30 days	20	Impairment in motor performance and cognition, a decrease in brain GSH and SOD.	[32]

Table 1.
Dose and time dependent exposure of acrylamide in rats.

discussed to understand the dose-effect of ACR includes 0.5–50 mg/kg [1, 7, 31–36] (**Table 1**). In rats, the NOAEL for ACR induced neurotoxicity was 0.5 mg/kg body weight/day and the LOAEL was 2 mg/kg-day in F344 male rats for the most sensitive effect (microscopic nerve alterations).

5. Role of oxidative stress in acrylamide induced neurotoxicity

The principle mechanism of ACR neurotoxicity is unknown, but some studies have linked it with the reduction in antioxidative capacity and inflammatory responses [7]. Oxidative stress which occurs due to the imbalance between the production and the removal of reactive oxygen species (ROS), free radicals and antioxidants, is evident in neurological disorders like Alzheimer's disease (AD), Huntington's disease (HD) and Parkinson's disease (PD), ataxia, peripheral neuropathy, amyotrophic lateral sclerosis (ALS) and multiple sclerosis (MS) [37–40]. Antioxidant enzymes like superoxide dismutase (SOD), catalase (CAT), glutathione S-transferase (GST), glutathione peroxidase (GSH-Px) neutralise the effects of free radicals but due to the pro-oxidant effect of ACR, the levels of these enzyme are decreased resulting in an imbalance between the production and removal of free radicals by elevating the oxidative stress markers like ROS and thereby inducing lipid peroxidation (LPO) [33, 41–43] (**Figure 1**). Like many other xenobiotic compounds, ACR is an electrophile that can interact with nucleophiles containing specific residues [43]. It reacts with molecules that consist of bisulfide (SH), azanide (NH2) or hydroxide (OH). Glutathione (GSH) is a thiol that is well known for its free radical and ROS scavenging property [7, 33, 34]. Previous research has established that rats administered with ACR have a significant decrease of GSH in brain tissue when compared to the control group [33, 34, 44, 45]. The reduction in GSH levels results in an increase in levels of ROS that accumulates and induces oxidative stress. GSH is a nonenzymatic

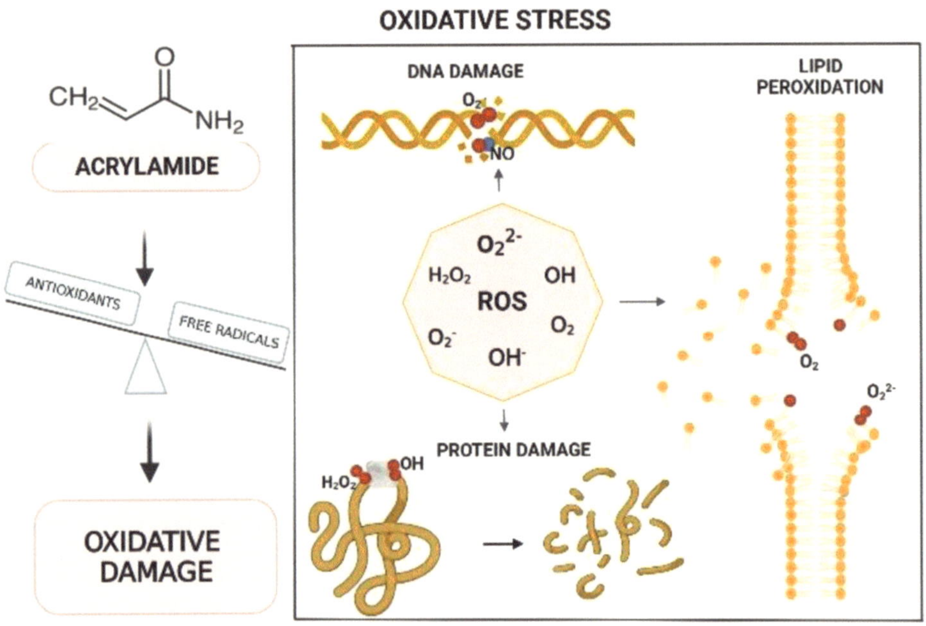

Figure 1.
Role of oxidative stress in ACR induced neurotoxicity.

antioxidant that also acts as a coenzyme for the peroxide decomposition enzyme GSH-Px [7]. Studies also suggest that LPO is an effect of low levels of GSH [46–48]. First-line defence enzymatic antioxidants like SOD, CAT and GSH-Px have reduced in ACR-treated rats [33, 42, 45, 47]. GST is an antioxidant enzyme used to maintain the free radical balance. ACR was shown to enhance GST activity, suggesting an increase in the synthesis of S-conjugates between ACR and GSH [43, 46]. In contrast, a study by [47] reported a decrease in GST activity while measuring ACR-treated rat brains. As a result of the changes in levels and activity of various antioxidant enzymes and molecules, the concentration of total oxidants and antioxidants are increased and decreased respectively [49]. LPO, protein damage and DNA damage are biomarkers for oxidative stress in neurodegenerative diseases [39, 50].

5.1 Lipid peroxidation

LPO is the degradation of lipids by free radicals and is assessed by measuring the levels of its marker, malondialdehyde (MDA). Thiobarbituric acid reactive substance (TBARS) assay is used to measure MDA, where it reacts with TBA and produces a pink-coloured complex [51]. MDA is formed by the free radicals generated from LPO and causes protein oxidation. Several lines of evidence suggest that exposing rats to ACR have high contents of MDA in the brain when compared to control groups [33, 40, 44, 47].

5.2 Protein degradation

Protein oxidation is considered as the damage of proteins. To understand the level of extent of this effect, the protein carbonyl content is the marker to understand oxidative damage in rats that have been treated with ACR. The Dinitrophenylhydrazine (DNPH) assay is used to measure the levels of protein-hydrazone to quantify the protein carbonyl content [42].

5.3 DNA damage

Since GA binds to the DNA and causes detrimental effects, it is important to understand the genotoxicity of ACR on rats. A commonly used DNA damage marker for oxidative stress is 8-OHdG, which is quantified by ELISA kits [52]. The principle mechanism of ACR-induced neurotoxicity is widely accepted to be apoptosis induced by ACR in rats [33]. ROS induces cell death via apoptotic mechanisms that are either non-physiological or controlled [53]. Telomerase reverse transcriptase (TERT) is an apoptosis-related molecule and is influenced by oxidative stress because of its anti-apoptotic effect. When ACR was administered to rats TERT associated mRNA and protein expression was downregulated in the rat brain [33, 54]. The sensitivity of cells to various apoptotic stimuli is determined by the ratio of anti-apoptotic protein B-cell lymphoma 2 (Bcl2) to pro-cell death proteins such as Bax and Bad, these ratios and the relative density of caspase-3 and caspase-9 is higher in ACR treated rats [37, 41, 55]. Proteins involved in apoptosis signalling pathways and cellular functions are also influenced by the presence of ACR. An appropriate balance must be maintained within the mitogen activated protein kinases (MAPKs) for regulating apoptosis. But, when ACR is induced, due to excessive ROS production a reduction in P-ERK/ERK ratio and elevation in the P-JNK/JNK and P-P38/P38 is observed, this causes mitochondrial dysfunction [1, 41].

6. Mechanisms underlying neurodegenerative diseases

The main neurotoxic consequences of ACR are peripheral nervous system (PNS) degradation and degeneration in a brain area related to learning and memory function. Drowsiness, cerebellar ataxia, muscle atrophy, dysarthria, and sensory or motor peripheral polyneuropathy are common clinical symptoms [56, 57]. ACR-induced neurotoxicity is associated with symptoms like ataxia, hindfoot splay, skeletal muscle weakness, and numbness of the hands and feet [1, 58]. ACR-induced neurodegenerative diseases have been shown in various studies to be mediated by axon and medullary sheath destruction in the PNS [59]. Distal axon swelling and degeneration are the key pathological features of ACR exposure [56, 60]. Even though recent research studies report that ACR-induced neurotoxicity and neurodegenerative effects in humans and experimental animals are mediated by nerve terminal and axonal damage, the exact underlying mechanism remains unclear [58].

6.1 Acrylamide induced dopaminergic neuronal loss in rat striatum

Recent research has revealed that ACR-induced locomotor abnormalities and neurotoxicity are comparable to the effects seen in PD, as ACR can cause key parkinsonian pathology such as α- synuclein aggregation [61]. The prominent hallmark of PD is the depletion of monoamine neurotransmitters (NTs) known as dopamine (DA) and its associated loss of dopaminergic A9 neurons in the substantia nigra pars compacta and striatum [58, 62, 63]. Motor control, cognitive decline, muscular stiffness, body posture instability, and movement difficulties are symptoms associated with the loss of dopaminergic A9 neurons [62]. DA, a kind of catecholamine, is a NT that governs important functions like cognition, motor control, emotion, and neuroendocrine activity [1, 7, 63]. A massive proportion of DA-carrying nigrostriatal neurons can be found in the striatum, which is the largest integral processing unit present in the basal ganglia [1, 63]. Tyrosine hydroxylase (TH) is a rate-limiting enzyme that is accountable for the synthesis of DA. TH helps to convert tyrosine into 3,4-dihydroxyphenylalanine (DOPA). DOPA is further converted to DA by the action of the enzyme, aromatic amino acid decarboxylase. Cells that are TH-positive are represented as dopaminergic neurons [1, 63, 64]. ACR is most commonly administered to rats either through their oral gavage or through intraperitoneal injection. When ACR is injected in this form, it gets metabolised into GA because of chromosome P450-2E1 present in the liver microsomes. DNA adducts that are formed as a result of the interaction between GA and DNA are responsible for provoking modxicity and carcinogenicity. Since ACR-induced neurotoxicity is strongly associated with the monomer of ACR itself, intracerebroventricular injection aids in transmitting the ACR to the neurons without resulting in the formation of GA. Studies have reported that rats treated with ACR through intracerebroventricular injection have shown a serious decline in the protein expression of TH and the number of TH-positive cells belonging to the striatum [1].

6.2 Acrylamide induced neuronal apoptosis in the rat striatum

Neuronal apoptosis results in the death of neuronal cells gradually leading to the development of neurodegenerative diseases. Neuronal apoptosis, a definite form of cell death, has a pivotal function in ACR-induced neurotoxicity in rats [65]. Studies have reported that ACR can result in neuronal apoptosis of the striatum. Nissl body is a chromatophilic substance that is very specific and is found in the cytoplasm of neuronal dendrites. Besides protein synthesis, Nissl bodies are crucial for brain functions like memory and learning. Since protein synthesis is essential for proper neuronal function, the presence of the Nissl body is indispensable. Rats treated with ACR reported the presence of pyknotic nuclei and the disappearance of Nissl substance in the striatal neurons. Striatal neurons treated with ACR also appeared swollen with decreased cellular integrity and exhibited an irregular arrangement [60]. Terminal deoxynucleotidyl transferase-mediated dUTP nick end labelling (TUNEL) staining is generally performed to investigate the loss of neurons due to apoptosis. Recent studies have reported that ACR treatment in rats has significantly raised the levels of TUNEL-positive cells in the rat striatum, which suggests that ACR exposure can result in striatal dopaminergic neuronal apoptosis. Hence, these kinds of studies suggest that ACR can also be a significant environmental risk factor for diseases like PD [1].

7. Therapeutic agents against neurotoxicity of acrylamide

To be able to mitigate the neurotoxic effects of ACR, therapeutic agents of different types are used at different doses. Phytoconstituents have been widely studied for amelioration of neurotoxicity in rats but there are adequate studies on dietary supplements, drugs and probiotics (**Figure 2**).

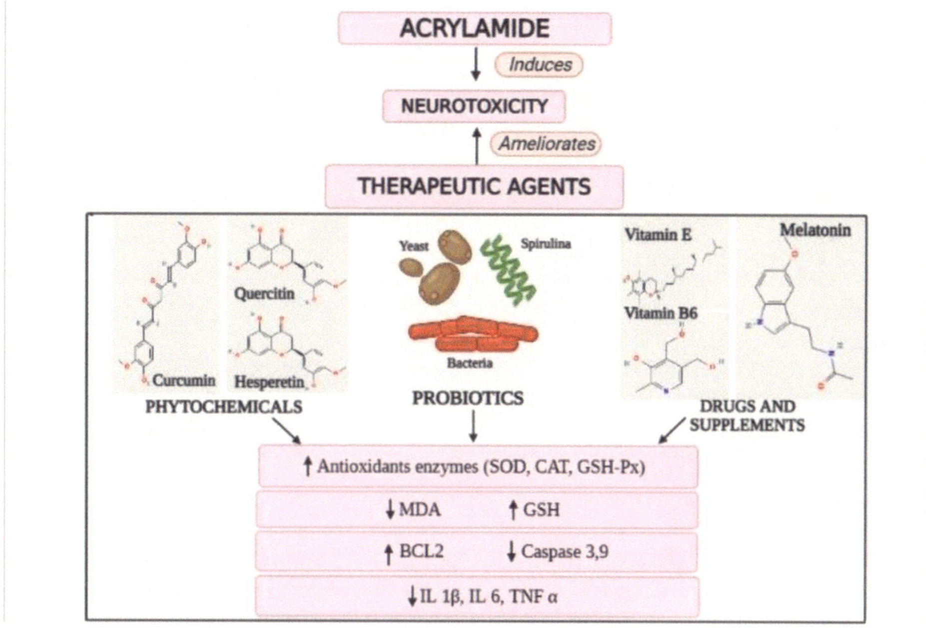

Figure 2.
Effects of therapeutic agents on neurotoxicity caused by acrylamide.

7.1 Phytochemicals

Few phytochemicals are thymoquinone, curcumin and quercetin. The anti-apoptotic property of thymoquinone plays a crucial role in attenuating the toxicity induced by ACR in rats by mitigating oxidative stress, reducing Bax/Bcl ratio, maintaining the integrity of the blood brain barrier (BBB), decreasing the level of caspase 3 and 9 and reducing glial fibrillary acidic protein (GFAP) content which indicates astrocyte damage [41, 66]. Curcumin increased the number of TERT positive cells and decreased the number of TUNEL positive cells in the cortex of ACR treated rats. Additionally, curcumin can also cross the BBB and alleviate spatial memory damage induced by ACR [33, 67]. Quercetin enhanced DA and serotonin levels, reduced biomarkers of oxidative stress, restored acetylcholinesterase (AChE) activity in ACR-treated rats. It can move across the BBB and exhibit its therapeutic efficiency [36, 68, 69]. Other compounds like metformin, minocycline and zolpidem also show similar therapeutic effects to dietary supplements when administered to ACR-treated rats [70–72] (**Table 2**).

7.2 Drugs and supplements

Vitamins have shown therapeutic effects when administered to ACR-induced rats by ameliorating their toxic effects. Vitamins like vitamin E, vitamin F, vitamin C and vitamin B6 have been studied for their ameliorative property on rats influenced by ACR neurotoxicity [37, 74, 75]. They are widely known for their powerful anti-oxidative property and are also used as positive control groups while evaluating the potential of other therapeutic agents against ACR toxicity in rats [41, 80]. Vitamin E is phospholipid soluble and a neuroprotective antioxidant. It elevated brain-derived

Therapeutic agents	Methods of exposure	Time exposure	Dose		References
			ACR (mg/kg)	Agent	
Thymoquinone	Primary treatment with Agent and followed by a concomitant treatment (ACR + AGENT)	11 days	50	2.5, 5, 10 mg/kg	[41, 66]
	Concomitant (ACR + AGENT)	11 days	50	2.5, 5, 10 mg/kg	
Curcumin	ACR + AGENT	4 weeks	40	50, 100 mg/kg	[33]
	Concomitant (ACR + AGENT)	7 weeks	10	90 mg/kg	[67]
Quercetin	Agent followed by ACR	5 days	50	10 mg/kg	[68]
	ACR followed by Agent	10 days	38.27	5, 10, 20, 40 mg/kg	[36]
	ACR followed by Agent	30 days	20	25, 50 mg/kg	[69]
Vitamin E	Concomitant (ACR + AGENT)	20 days	5	100 mg/kg	[49, 73]
	Concomitant (ACR + AGENT)	28 days	20	50 IU/kg	
	ACR followed by Agent	42 days	20	50 IU/kg	
Vitamin C	ACR + Agent	21 days	10	200 mg/kg	[74]
Vitamin F	ACR followed by agent	13 days	38.27	5, 10, 20, 40 mg/kg	[75]
Omega-3 Fish oil	Concomitant (ACR + Agent)	8 weeks	45	200 mg/kg	[65, 76, 77]
	Agent followed by ACR	30 days	30	0.5 ml/kg	
Melatonin	ACR + Agent	21 days	50	10 mg/kg	[34, 78, 79]

Table 2.
Therapeutic agents that attenuate acrylamide neurotoxicity in rats by different methods of exposure at various doses and times.

neurotrophic factor levels and lessened oxidative stress through its sweeper effect and removed free radicals in the brain tissue of fetal rats [49]. It also attenuated inflammation, apoptosis and behavioural neurotoxic effects in rats [37, 73]. Linoleic Acid (LA) is an essential omega-6 fatty acid with antioxidative, anti-inflammatory and neuroprotective effects [75]. LA improved ACR oxidative effects by restoring the activities of antioxidant enzymes, reducing the generation of free radicals, preventing LPO and obstructing genotoxic damage by reducing GA. AchE activity was also ameliorated by restoring vacuolization loss by pyramidal cells and Purkinje cells [75]. Vitamin B6 was also able to attenuate the intensity of ACR effects by increasing the availability of energy to the neurons [81]. When administered to pregnant rats vitamin C lessens the effects in white matter volume, the volume of the cerebellar cortex, molecular and granular layer volume and cerebellum damage [74]. Omega-3 fatty acids have also been studied as a therapeutic agent that can attenuate neurotoxicity caused by ACR in rats. Fish oil was able to reduce the neurotoxic effects evoked by ACR in rats. It restored oxidative stress by improving MDA, GSH, LPO, protein carbonyl content, free radicals and antioxidant status [65, 76]. Omega-3 Polyunsaturated fatty acids (PUFAs) regulate neurotransmission by modulating the activity of NTs. They also attenuate apoptosis by increasing anti-apoptotic BCL-2, expressing Hsp27 and inducing oligodendrogenesis [77]. Fish oil mitigates inflammation and astrogliosis by reducing inflammatory cytokines and GFAP positive cells [76]. Melatonin (MT) alleviates DNA damage, levels of MDA, SOD, GSH-Px, GSH and nucleus concentration [34, 78]. It relieved weight loss and gait abnormality. MT shows an increase in the levels of brain NTs and a reduction in AchE activity, serum tumour necrosis factor (TNF)—α and cortical amyloid protein levels [79]. MT treatment restored ACR evoked oxidative stress by down-regulating Nrf2, nuclear factor kappa B (NF-kB) and Kelch-like ECH-associated protein 1 (Keap-1) activity (**Table 2**) [78].

7.3 Probiotics

Probiotics maintain the intestinal barrier by increasing the expression of tight junction proteins, they can even reduce toxic substance absorption in the gut and enhance angiogenic activities in the central nervous system (CNS) [45, 82]. Lactobacillus plantarum (*L. plantarum*) ATCC8014 was studied to understand how it attenuates ACR induced toxicity in rats. This strain of Lactic acid bacteria is chosen because it has a high absorption rate of ACR and antioxidant capacity [45]. At high doses, *L. plantarum* ATCC8014 increased the body's weight growth in ACR-treated rats. Administration of *L. plantarum* ATCC8014 improved nerve tissue damage and elevated the antioxidant capacity of nerve tissue by preventing attacks because of its effective capacity to scavenge and reduce free radicals. Similarly, probiotic *Enterococcus faecium* NCIM 5593 showed improved protection against the neurodegenerative changes due to oxidative damage in adult mice [83]. Additionally, it is important to consider the effect of prebiotics while considering attenuating properties of probiotics. Prebiotics are fibre-rich foods that promote the growth of probiotic microorganisms in the gastrointestinal tract. The study by [84] explored the effects of oral prebiotic supplements containing fructo- and xylo-oligosaccharides on pregnant rats exposed to ACR. It was found that AChE activity was restored and DA levels increased in the cortex of rats after administering prebiotics. Also, spirulina is a prebiotic obtained from the blue-green algae, *Arthrospira platensis.* It is the dried biomass of this cyanobacterium. It contains many beneficial compounds like proteins, vitamins, phytochemicals, etc. [85]. Spirulina is also known for its antioxidant properties owing to the presence of compounds like C-phycocyanin, and ß-carotene [86]. The study by [86] showed that spirulina was effective in ameliorating the toxicity induced by ACR in rats in a dose-dependent manner. Administration of spirulina enhanced the antioxidant activity and reduced the levels of TNF-α, IL-1β, and IL-6 in the serum. This shows the enormous scope of studies with regard to probiotics and prebiotics and their ability to attenuate ACR-induced neurotoxicity. Further studies are required to understand the protective role of probiotics and prebiotics.

8. Management and mitigation of acrylamide and its by-product wastes

In many countries, the current standard for ACR concentration in drinking water is 0.25 g/litre. It is advised to maintain the level of ACR monomer at 0.05 percent in PAM used in wastewater treatment. A wide range of microbes can degrade ACR, but there exists a latent period before this occurs. However, in regions with low microbial activity, ACR may remain in the environment for days, weeks or even months. ACR contamination also occurs during sewage treatment. This can be mitigated by chemically decontaminating ACR containing effluents [87]. The limitation set by Food and Drug Administration is of 0.2 percent (2 g/kg) monomer in PAM for use in paper or food or cardboard. In the Federal Republic of Germany, the amount of PAM in packaged foods is regulated to 0.3 percent (3 g/kg) and the amount of residual ACR monomer is limited to 0.2 percent (2 g/kg). Various methods such as addition of divalent cations, replacement of reducing sugars with sucrose or addition of organic acids, addition of calcium salts, using glycine to dilute the asparagine level, reduction in the free asparagine concentration by asparaginase or substitution of ammonium salts with baking powder, are suggested in recent years to mitigate the formation of ACR in heat processed foods [87].

The study by [88] showed a decline in ACR content in baked corn chips and French fries by pre-treating the potato cuts with citric acid solution prior to frying. The citric acid solution was able to lower the pH and leach out the asparagine and reduce sugar from the potato cuts. Aiswarya and Baskar [89] showed that the pre-treatment of potato with asparaginase prior to frying was effective in reducing the ACR content in fried potato chips. The effects of NaCl and citric acid combined with asparaginase was also studied and it was found that the use of NaCl + asparaginase and citric acid + asparaginase was effective in reducing ACR levels. To prevent workers from absorbing more than 0.012 mg/kg body weight per day during their occupational exposure, preventive measures such as enclosing production activities and wearing protective garments should be implemented. In the workroom, the concentration of ACR in the air should not exceed 0.1 mg/m3. To avoid inhaling ACR, ventilated face masks can be worn. It's likely that the underlying neurological disease and/or the administration of neuroactive treatments modify human sensitivity to ACR, but no particular recommendations could be given until there is evidence [87].

9. Conclusions

Many studies have identified the potential health risks of ACR and the ambiguity of the mechanisms underlying ACR induced neurotoxicity has gained interest. Current toxicological studies are insufficient to indicate that ACR amounts consumed in the normal diet are likely to result in adverse human health effects. ACR is considered to be a potential health hazard that can impact toxicity to humans. An overview of their occurrence and effects have been comprehended in this review chapter. The importance of oxidative stress, dose and time variations in exposed rat models is being used to comprehend the mechanisms and the neurotoxic effects induced by ACR in rat models. ACR, a toxic neurotoxin is associated with the pathogenesis of various neurodegenerative diseases and their effects on neuronal apoptosis are analysed. Various therapeutic agents against ACR induced neurotoxicity have been analysed to understand their ameliorative effect. This review would give an overall insight on the toxicological effects of acrylamide and provides a comprehensive approach about the recent findings on how to mitigate the formation of acrylamide by using effective therapeutic strategies. More research at the cellular level will aid in the identification of early biomarkers that can be utilised to detect, avoid or mitigate the effects of ACR induced neurotoxicity.

Acronyms and abbreviations

ACR: acrylamide; NOAEL: no-observed-adverse-effect limit; LOAEL: lowest observed adverse effect level; BMDL: benchmark dose lower confidence limit; PAM: polyacrylamide; CYPs: Cytochrome; GA: glycidamide; CNS: central nervous system; PNS: peripheral nervous system; MT: melatonin; GSH: glutathione; SOD: superoxide dismutase; CAT: catalase; SH: bisulfide; NH2: azanide; OH: hydroxide; NT: neurotransmitters; TBA: thiobarbituric acid; LA: linolenic acid; PUFA: polyunsaturated fatty acids; GST: glutathione S-transferase; MDA: malondialdehyde; DNPH: dinitrophenylhydrazine; TERT: telomerase reverse transcriptase; ROS: reactive oxygen species; GFAP: glial fibrillary acidic protein; TBARS: thiobarbituric acid reactive substance; TUNEL: terminal deoxynucleotidyl transferase-mediated dUTP nick end labelling; AChE: acetylcholinesterase; DA: dopamine; Bcl2: B-cell lymphoma 2; MAPK: mitogen-activated protein kinase; TNF: tumour necrosis factor; LPO: lipid peroxidation; NF-kB: nuclear factor kappa B; Keap-1: Kelch-like ECH-associated protein 1; TH: tyrosine hydroxylase; DOPA: 3,4-dihydroxyphenylalanine; BBB: blood brain barrier; AD: Alzheimer's disease; PD: Parkinson's disease; HD: Huntington's disease; ALS: amyotrophic lateral sclerosis; and MS: multiple sclerosis.

Author details

Prathyusha Cota, Sayantani Saha, Shailvi Tewari, Abhirami Sasikumar, M. Yashwant Saran, Swetha Senthilkumar and Sahabudeen Sheik Mohideen*
Department of Biotechnology, School of Bio-Engineering, SRM Institute of Science and Technology, Chennai, India

*Address all correspondence to: sahabuds@srmist.edu.in

References

[1] Yan D, Pan X, Yao J, Wang D, Wu X, Chen X, et al. MAPKs and NF-κB-mediated acrylamide-induced neuropathy in rat striatum and human neuroblastoma cells SY5Y. Journal of Cellular Biochemistry. 2019;**120**(3):3898-3910

[2] Dasari S, Ganjayi MS, Meriga B. Glutathione S-transferase is a good biomarker in acrylamide induced neurotoxicity and genotoxicity. Interdisciplinary Toxicology. 2018;**11**(2): 115-121

[3] Bušová M, Bencko V, Laktičová KV, Holcátová I, Vargová M. Risk of exposure to acrylamide. Central European Journal of Public Health. 2020;**28**(Suppl):S43-S46

[4] Ubaoji KI, Orji VU. A review on acrylamide in foods: Sources and implications to health. Journal of African Studies. 2015;**4**(1):1-12

[5] Zamani E, Shokrzadeh M, Fallah M, Shaki F. A review of acrylamide toxicity and its mechanism. Pharmaceutical and Biomedical Research. 2017;**3**(1):1-7

[6] EFSA Panel on Contaminants in the Food Chain (CONTAM). Scientific Opinion on acrylamide in food. EFSA Journal. 2015;**13**(6):4104. pp. 321. DOI: 10.2903/j.efsa.2015.4104

[7] Farouk SM, Gad FA, Almeer R, Abdel-Daim MM, Emam MA. Exploring the possible neuroprotective and antioxidant potency of lycopene against acrylamide-induced neurotoxicity in rats' brain. Biomedicine & Pharmacotherapy. 2021;**138**:111458

[8] Food and Drug Administration. Acrylamide|FDA. 2019. Available from: https://www.fda.gov/food/chemical-contaminants-food/acrylamide [Accessed: January 8, 2022]

[9] Moorman WJ, Reutman SS, Shaw PB, Blade LM, Marlow D, Vesper H, et al. Occupational exposure to acrylamide in closed system production plants: Air levels and biomonitoring. Journal of Toxicology and Environmental Health—Part A: Current Issues. 2012;**75**:100-111

[10] Junqua G, Spinelli S, Gonzalez C. Occurrence and fate of acrylamide in water-recycling systems and sludge in aggregate industries. Environmental Science and Pollution Research. 2015;**22**(9):6452-6460

[11] Kusnin N, Syed MA, Ahmad SA. Toxicity, pollution and biodegradation of acrylamide—A mini review. Journal of Biochemistry, Microbiology and Biotechnology. 2015;**3**(2):6-12

[12] Weideborg M, Källqvist T, Ødegård KE, Sverdrup LE, Vik EA. Environmental risk assessment of acrylamide and methylolacrylamide from a grouting agent used in the tunnel construction of romeriksporten, Norway. Water Research. 2001;**35**(11):2645-2652

[13] Tepe Y, Çebi A. Acrylamide in environmental water: A review on sources, exposure, and public health risks. Exposure and Health. 2019;**11**(1):3-12

[14] Yue Z, Tian E, Chen Y, Luo L, Yang L, He L, et al. The adverse effects of acrylamide exposure on the early development of marine medaka (Oryzias melastigma) and its mechanisms. Marine Pollution Bulletin. 2021 Feb 1;**163**:111875

[15] Friedman M. Acrylamide: Inhibition of formation in processed food and

mitigation of toxicity in cells, animals, and humans. Food & Function. 2015;**6**(6):1752-1772

[16] Pruser KN, Flynn NE. Acrylamide in health and disease. Frontiers in Bioscience. 2011;**3**:41-51

[17] Mojska H, Gielecińska I, Cendrowski A. Acrylamide content in cigarette mainstream smoke and estimation of exposure to acrylamide from tobacco smoke in Poland. Annals of Agricultural and Environmental Medicine. 2016;**23**(3):456-461

[18] Schettgen T, Rossbach B, Kütting B, Letzel S, Drexler H, Angerer J. Determination of haemoglobin adducts of acrylamide and glycidamide in smoking and non-smoking persons of the general population. International Journal of Hygiene and Environmental Health. 2004;**207**(6):531-539

[19] Kim H, Lee SG, Rhie J. Dermal and neural toxicity caused by acrylamide exposure in two Korean grouting workers: a case report. Annals of Occupational and Environmental Medicine. 2017 Dec;**29**(1):1-6

[20] Pantusa VP, Stock TH, Morandi MT, Harrist RB, Afshar M. Inhalation exposures to acrylamide in biomedical laboratories. AIHA Journal. 2010;**63**(4):468-473. DOI: 10.1080/15428110208984735

[21] Rifai L, Saleh FA. A review on acrylamide in food: Occurrence, toxicity, and mitigation strategies. International Journal of Toxicology. 2020;**39**(2):93-102

[22] Kumar J, Das S, Teoh SL. Dietary acrylamide and the risks of developing cancer: Facts to ponder. Frontiers in Nutrition. 2018;**0**:14

[23] Shipp A, Lawrence G, Gentry R, McDonald T, Bartow H, Bounds J, et al. Acrylamide: Review of toxicity data and dose-response analyses for cancer and noncancer effects. Critical Reviews in Toxicology. 2006;**36**(6-7):481-608

[24] Segerbmck D, Calleman CJ, Schroeder JL, Costa LG, Faustman EM. Formation of N-7-(2-carbamoyl-2-hydroxyethyl)guanine in DNA of the mouse and the rat following intraperitoneal administration of [14 C]acrylamide. Carcinogenesis. 1995;**16**(5):1161-1165

[25] Gamboa da Costa G, Churchwell MI, Hamilton LP, Von Tungeln LS, Beland FA, Marques MM, et al. DNA adduct formation from acrylamide via conversion to glycidamide in adult and neonatal mice. Chemical research in toxicology. 20 Oct 2003;**16**(10):1328-1337

[26] Calleman CJ, Bergmark E, Costa LG. Acrylamide is metabolized to glycidamide in the rat: evidence from hemoglobin adduct formation. Chemical Research in Toxicology. Sep 1990;**3**(5):406-412

[27] Arinc E. Effects of diabetes on rabbit kidney and lung CYP2E1 and CYP2B4 expression and drug metabolism and potentiation of carcinogenic activity of N-nitrosodimethylamine in kidney and lung. Food and Chemical Toxicology. 2007;**45**:107-118

[28] Paulsson B, Kotova N, Grawé J, Henderson A, Granath F, Golding B, et al. Induction of micronuclei in mouse and rat by glycidamide, genotoxic metabolite of acrylamide. Mutation Research/Genetic Toxicology and Environmental Mutagenesis. 2003;**535**(1):15-24

[29] Besaratinia A, Pfeifer GP. Weak yet distinct mutagenicity of acrylamide in mammalian cells. Journal of the National Cancer Institute. 2003;**95**(12):889-896

[30] Erdemli ME, Doğan Z, Çiğremiş Y, Akgöz M, Altinöz E, Geçer M, et al. Amelioration of subchronic acrylamide toxicity in large intestine of rats by organic dried apricot intake. Turkish Journal of Biology. 2015;**39**(6):872-878

[31] Stokinger HE. In the field of toxicology. American Industrial Hygiene Association Quarterly. 1956;**17**(3):340-344

[32] Goudarzi M, Mombeini MA, Fatemi I, Aminzadeh A, Kalantari H, Nesari A, et al. Neuroprotective effects of ellagic acid against acrylamide-induced neurotoxicity in rats. Neurological Research. 2019;**41**(5):419-428

[33] Guo J, Cao X, Hu X, Li S, Wang J. The anti-apoptotic, antioxidant and anti-inflammatory effects of curcumin on acrylamide-induced neurotoxicity in rats. BMC Pharmacology and Toxicology. Dec 2020;**21**(1):1-10

[34] Pan X, Zhu L, Lu H, Wang D, Lu Q, Yan H. Melatonin attenuates oxidative damage induced by acrylamide in vitro and in vivo. Oxidative Medicine and Cellular Longevity. 2015;**2015**:1-12. DOI: 10.1155/2015/703709

[35] Tian SM, Ma YX, Shi J, Lou TY, Liu SS, Li GY. Acrylamide neurotoxicity on the cerebrum of weaning rats. Neural Regeneration Research. 2015;**10**(6):938-943

[36] Uthra C, Shrivastava S, Jaswal A, Sinha N, Reshi MS, Shukla S. Therapeutic potential of quercetin against acrylamide induced toxicity in rats. Biomedicine & Pharmacotherapy. 2017;**86**:705-714

[37] Bainmahfouz FR, Ali SS, Al-Shali RA, El-Shitany NA. Vitamin E and 5-amino salicylic acid ameliorates acrylamide-induced peripheral neuropathy by inhibiting caspase-3 and inducible nitric oxide synthase immunoexpression. Journal of Chemical Neuroanatomy. 2021 Apr 1;**113**:101935

[38] Abdel-Daim MM, Abo El-Ela FI, Alshahrani FK, Bin-Jumah M, Al-Zharani M, Almutairi B, et al. Protective effects of thymoquinone against acrylamide-induced liver, kidney and brain oxidative damage in rats. Environmental Science and Pollution Research. Oct 2020;**27**(30):37709-37717.

[39] Santhanasabapathy R, Vasudevan S, Anupriya K, Pabitha R, Sudhandiran G. Farnesol quells oxidative stress, reactive gliosis and inflammation during acrylamide-induced neurotoxicity: Behavioral and biochemical evidence. Neuroscience. 2015;**308**:212-227. DOI: 10.1016/j.neuroscience.2015.08.067

[40] Mehri S, Abnous K, Khooei A, Mousavi SH, Shariaty VM, Hosseinzadeh H. Crocin reduced acrylamide-induced neurotoxicity in wistar rat through inhibition of oxidative stress. Iranian Journal of Basic Medical Sciences. 2015;**18**(9):902-908

[41] Tabeshpour J, Mehri S, Abnous K, Hosseinzadeh H. Role of oxidative stress, MAPKinase and apoptosis pathways in the protective effects of thymoquinone against acrylamide-induced central nervous system toxicity in rat. Neurochemical Research. 2020;**45**(2):254-267. DOI: 10.1007/s11064-019-02908-z

[42] Abo-El-Matty DM, Rizk MZ, Aly HF, Abd-Alla HI, Saleh SM, Younis EA. Role of fruit waste and flavanones-loaded silica nanoparticles in ameliorating oxidative stress and histological changes in rat brain induced by acrylamide. Journal of Materials and Environmental Science. 2018;**9**(6):1817-1828

[43] Yousef MI, El-Demerdash FM. Acrylamide-induced oxidative stress

and biochemical perturbations in rats. Toxicology. 2006;**219**(1-3):133-141

[44] Gur C, Kandemir FM, Darendelioglu E, Caglayan C, Kucukler S, Kandemir O, et al. Morin protects against acrylamide-induced neurotoxicity in rats: An investigation into different signal pathways. Environmental Science and Pollution Research. 2021;**28**(36):49808-49819

[45] Zhao S, Zhao X, Liu Q, Jiang Y, Li Y, Feng W, Xu H, Shao M. Protective effect of Lactobacillus plantarum ATCC8014 on acrylamide-induced oxidative damage in rats. Applied Biological Chemistry. Dec 2020;**63**(1):1-4

[46] Ghareeb DA, Khalil AA, Elbassoumy AM, Hussien HM, Abo-Sraiaa MM. Ameliorated effects of garlic (Allium sativum) on biomarkers of subchronic acrylamide hepatotoxicity and brain toxicity in rats. Toxicological and Environmental Chemistry. 2010;**92**(7):1357-1372. DOI: 10.1080/02772240903348187

[47] Lebda MA, Gad SB, Rashed RR. The effect of lipoic acid on acrylamide-induced neuropathy in rats with reference to biochemical, hematological, and behavioral alterations. Pharmaceutical Biology. 2015;**53**(8):1207-1213. DOI: 10.3109/13880209.2014.970288

[48] Prasad SN. Muralidhara. Neuroprotective efficacy of eugenol and isoeugenol in acrylamide-induced neuropathy in rats: Behavioral and biochemical evidence. Neurochemical Research. 2013;**38**(2):330-345

[49] Erdemli ME, Turkoz Y, Altinoz E, Elibol E, Dogan Z. Investigation of the effects of acrylamide applied during pregnancy on fetal brain development in rats and protective role of the Vitamin E. Human & Experimental Toxicology. 2016;**35**(12):1337-1344

[50] Salim S. Minireviews oxidative stress and the central nervous system. Journal of Pharmacology and Experimental Therapeutics. 2017;**360**:201-205. DOI: 10.1124/jpet.116.237503

[51] Mehri S, Meshki MA, Hosseinzadeh H. Linalool as a neuroprotective agent against acrylamide-induced neurotoxicity in wistar rats. Drug and Chemical Toxicology. 2015;**38**(2):162-166

[52] Alturfan AA, Tozan-Beceren A, Şehirli AÖ, Demiralp E, Şener G, Omurtag GZ. Resveratrol ameliorates oxidative DNA damage and protects against acrylamide-induced oxidative stress in rats. Molecular Biology Reports. 2012;**39**(4):4589-4596. DOI: 10.1007/s11033-011-1249-5

[53] JG L, YS W, CC C. Acrylamide-induced apoptosis in rat primary astrocytes and human astrocytoma cell lines. Toxicology In Vitro. 2014;**28**(4):562-570

[54] Wang J, Zhang MY, Xu SQ, Cheng J, Yu ZJ, Hu XM. Down-regulation of telomerase reverse transcriptase-related anti-apoptotic function in a rat model of acrylamide induced neurobehavioral deficits. Biotechnic & Histochemistry. 2018;**93**(7):512-518

[55] Ghasemzadeh Rahbardar M, Hemadeh B, Razavi BM, Eisvand F, Hosseinzadeh H. Effect of carnosic acid on acrylamide induced neurotoxicity: in vivo and in vitro experiments. Drug and Chemical Toxicology. 18 Nov 2020:1-8. DOI: 10.1080/01480545.2020.1845715

[56] X W, F Y, M E, C Z, G L, X Y, et al. Neuroprotective Effect of Calpeptin on Acrylamide-Induced Neuropathy

in Rats. Neurochemical Research. 2015;**40**(11):2325-2332

[57] Su B, Guan Q, Wang M, Liu N, Wei X, Wang S, et al. Calpeptin is neuroprotective against acrylamide-induced neuropathy in rats. Toxicology. 2018;**400-401**:1-8

[58] Zong C, Hasegawa R, Urushitani M, Zhang L, Nagashima D, Sakurai T, et al. Role of microglial activation and neuroinflammation in neurotoxicity of acrylamide in vivo and in vitro. Archives of Toxicology. 2019;**93**(7):2007-2019

[59] Sun G, Qu S, Wang S, Shao Y, Sun J. Taurine attenuates acrylamide-induced axonal and myelinated damage through the Akt/GSK3β-dependent pathway. International journal of immunopathology and pharmacology. Oct 2018;32:2058738418805322

[60] Lai SM, Gu ZT, Zhao MM, Li XX, Ma YX, Luo L, et al. Toxic effect of acrylamide on the development of hippocampal neurons of weaning rats. Neural Regeneration Research. 2017;**12**(10):1648-1654

[61] Sui X, Yang J, Zhang G, Yuan X, Li W, Long J, Luo Y, Li Y, Wang Y. NLRP3 inflammasome inhibition attenuates subacute neurotoxicity induced by acrylamide in vitro and in vivo. Toxicology. 28 Feb 2020;**432**:152392

[62] Raina A, Leite K, Guerin S, Mahajani SU, Chakrabarti KS, Voll D, Becker S, Griesinger C, Bähr M, Kügler S. Dopaminepromotestheneurodegenerative potential of β-synuclein. Journal of Neurochemistry. Mar 2021;**156**(5): 674-91

[63] Vecchio LM, Sullivan P, Dunn AR, Bermejo MK, Fu R, Masoud ST, et al. Enhanced tyrosine hydroxylase activity induces oxidative stress, causes accumulation of autotoxic catecholamine metabolites, and augments amphetamine effects in vivo. Journal of Neurochemistry. 2021;**158**(4):960-979

[64] Miyajima K, Kawamoto C, Hara S, Mori-Kojima M, Ohye T, Sumi-Ichinose C, et al. Tyrosine hydroxylase conditional KO mice reveal peripheral tissue-dependent differences in dopamine biosynthetic pathways. The Journal of Biological Chemistry. 2021;**296**:100544

[65] Lakshmi D, Gopinath K, Jayanthy G, Anjum S, Prakash D, Sudhandiran G. Ameliorating effect of fish oil on acrylamide induced oxidative stress and neuronal apoptosis in cerebral cortex. Neurochemical Research. 2012;**37**(9):1859-1867. DOI: 10.1007/s11064-012-0794-1

[66] Tabeshpour J, Mehri S, Abnous K, Hosseinzadeh H. Neuroprotective effects of thymoquinone in acrylamide-induced peripheral nervous system toxicity through MAPKinase and apoptosis pathways in rat. Neurochemical Research. 2019;**44**:1101-1112

[67] Yan D, Yao J, Liu Y, Zhang X, Wang Y, Chen X, et al. Tau hyperphosphorylation and P-CREB reduction are involved in acrylamide-induced spatial memory impairment: Suppression by curcumin. Brain, Behavior, and Immunity. 2018;**71**:66-80. DOI: 10.1016/j.bbi.2018.04.014

[68] Zargar S, Siddiqi NJ, Ansar S, Alsulaimani MS, El Ansary AK. Therapeutic role of quercetin on oxidative damage induced by acrylamide in rat brain. Pharmaceutical Biology. 2016;**54**(9):1763-1767. DOI: 10.3109/13880209.2015.1127977

[69] El-Beltagi HS, Ahmed MM. Assessment the protective role of quercetin on acrylamide-induced oxidative stress in rats. Journal of Food Biochemistry. 2016;**40**(6):715-723

[70] Radad K, El Amir Y, Al-Emam A, Al-Shraim M, Bin-Jaliah I, Krewenka C, et al. Minocycline protects against acrylamide-induced neurotoxicity and testicular damage in sprague-dawley rats. Journal of Toxicologic Pathology. 2020;**33**(2):87

[71] Oda SS. Metformin protects against experimental acrylamide neuropathy in rats. Drug Development Research. 2017;**78**(7):349-359

[72] sadat Yousefsani B, Akbarizadeh N, Pourahmad J. The antioxidant and neuroprotective effects of Zolpidem on acrylamide-induced neurotoxicity using Wistar rat primary neuronal cortical culture. Toxicology Reports. 2020;7:233-240

[73] Rahangadale S, Kurkure N, Prajapati B, Hedaoo V, Bhandarkar AG. Neuroprotective effect of vitamin e supplementation in wistar rat treated with acrylamide. Toxicology International. 2012;**19**(1):1-8

[74] Dortaj H, Yadegari M, Hosseini Sharif Abad M, Abbasi Sarcheshmeh A, Anvari M. Stereological method for assessing the effect of vitamin C administration on the reduction of acrylamide-induced neurotoxicity. Basic and Clinical Neuroscience. 2018;**9**(1): 27-34. DOI: 10.29252/nirp.bcn.9.1.27

[75] Shrivastava S, Nirala SK, Reshi MS, Shukla S, Sharma A, Uthra C. Protective efficacy of vitamin F against acrylamide induced toxicity: Studies on oxidative stress biomarkers. Open Biomarkers Journal. 2019;**9**(1):62-69

[76] Elblehi SS, El Euony OI, El-Sayed YS. Apoptosis and astrogliosis perturbations and expression of regulatory inflammatory factors and neurotransmitters in acrylamide-induced neurotoxicity under ω3 fatty acids protection in rats. Neurotoxicology. 2020;**76**:44-57. DOI: 10.1016/j.neuro.2019.10.004

[77] Imam RA, Gadallah HN. Acrylamide-induced adverse cerebellar changes in rats: Possible oligodendrogenic effect of omega 3 and green tea. Folia Morphologica. 2019;**78**(3):564-574

[78] Edres HA, Taha NM, Lebda MA, Elfeky MS. The potential neuroprotective effect of allicin and melatonin in acrylamide-induced brain damage in rats. Environmental Science and Pollution Research. Nov 2021;**28**(41):1-13. DOI: 10.1007/ s11356-021-14800-x

[79] Ghada ZA, Soliman GZAS. Protective Effect of Solanum nigrum, Vitamin C or Melatonin on the Toxic Effect of Acrylamide on Rats. IOSR Journal of Pharmacy and Biological Science. 2013;**5**(5):47-54

[80] Zhao S, Sun H, Liu Q, Shen Y, Jiang Y, Li Y, et al. Protective effect of seabuckthorn berry juice against acrylamide-induced oxidative damage in rats. Journal of Food Science. 2020;**85**(7):2245-2254. DOI: 10.1111/1750-3841.15313

[81] Loeb AL, Anderson RJ. Antagonism of acrylamide neurotoxicity by supplementation with vitamin B6. Neurotoxicology. 1981;**2**(4):625-633

[82] Seifati SM, Zaker E, Fesahat F, Zare F, Hekmatimoghaddam S. Modulatory effect of probiotics on proinflammatory cytokine levels in acrylamide-treated rats. Biochemistry Research International. 2021;**2021**:1-6.

[83] Divyashri G, Prapulla SG. Protective Effect of Probiotic Enterococcus faecium NCIM 5593 on Acrylamide Induced Neurotoxicity in Adult Mice. Journal of Probiotics and Health. 2017;**5**(1):1-11

[84] Krishna G. Oral supplements of combined fructo- and xylo-oligosaccharides during perinatal period significantly offsets acrylamide-induced oxidative impairments and neurotoxicity in rats. Journal of Physiology and Pharmacology. 2018;**69**(5):801-814

[85] Kordowska-Wiater M, Waśko A, Polak-Berecka M, Kubik-Komar A, Targoński Z. Spirulina enhances the viability of Lactobacillus rhamnosus E/N after freeze-drying in a protective medium of sucrose and lactulose. Letters in Applied Microbiology. 2011;**53**(1):79-83

[86] Bin-Jumah M, Abdel-Fattah A-FM, Saied EM, El-Seedi HR, Abdel-Daim MM. Acrylamide-induced peripheral neuropathy: Manifestations, mechanisms, and potential treatment modalities. Environmental Science and Pollution Research International. 2021;**28**(11):13031-13046

[87] Environmental Health Criteria 49 ACRYLAMIDE. World Heal Orgnization. 1985;1-85.

[88] Jung MY, Choi DS, Ju JW. A novel technique for limitation of acrylamide formation in fried and baked corn chips and in french fries. Journal of Food Science. 2003;**68**(4):1287-1290

[89] Aiswarya R, Baskar G. Enzymatic mitigation of acrylamide in fried potato chips using asparaginase from Aspergillus terreus. International Journal of Food Science and Technology. 2018;**53**(2):491-498

Section 2

Bioremediation of Hazardous Waste

Chapter 2

Bioremediation of Hazardous Wastes

Donald Tyoker Kukwa, Felicia Omolara Afolabi, Emmanuel Kweinor Tetteh, Ifeanyi Michael Smarte Anekwe and Maggie Chetty

Abstract

The remediation of the contaminated environment using the physical, thermal, or chemical methods has been criticized due to their high-cost implication, non-eco-friendly and inability to meet remediation objectives. Bioremediation offers the application of environmentally benign and cost-effective biological techniques for the remediation of contaminated sites. This chapter provides an overview of bioremediation technologies for the remediation of hazardous substances in the environment while highlighting the application of bioturbation as a promising bioremediation tool for the effective treatment of organic and inorganic contaminants. Given the success of bioremediation, most of these technologies are yet to be applied on a large scale which presents a drawback to this technique. Challenges and prospects for the effective application of bioremediation technologies were discussed.

Keywords: bioaugmentation, biostimulation, bioturbation, bioventing, genetically engineered microbes, phytoremediation

1. Introduction

The increase in human activities triggers environmental pollution through the generation and disposal of hazardous wastes in aquatic and terrestrial habitats [1]. Most of these pollutants include inorganic (heavy metals) and organic matter (polyaromatic hydrocarbons (PAH), petroleum hydrocarbons compounds (PHC)) which may cause negative effects on the ecosystem and possibly react with other abiotic factors that attribute to the effect on the structural arrangement of terrestrial and aquatic habitats [2].

In terms of the environment and ecology system, the proper and safe disposal of these hazardous wastes is a key priority for a sustainable ecosystem. This involves the use of various treatment procedures to clean up hazardous waste. For detoxifying heavy metals, radionuclide and organic polluted soils, physicochemical techniques such as filtration, precipitation, electrochemical treatment, soil washing and chelating, oxidation/reduction, ion exchange, reverse osmosis, and stabilization/

solidification have been employed. These environmental clean-up procedures have various disadvantages, including inefficiency, the need for a large number of chemical reagents, energy, and high cost, as well as the formation of secondary by-products [3].

Bioremediation is a cost-effective and environmentally tolerable technology that employs a biological process to reduce environmental risks caused by toxic substances and other hazardous pollutants. To treat polluted multiphase systems and sustain the native ecosystem, a combination of bioremediation techniques will be effective. The fundamental premise of bioremediation is to reduce contaminant solubility by adjusting pH, modifying redox processes, and adsorbing toxic substances from polluted sites [3]. Environmental remediation always requires human assistance to achieve effective remediation of contaminants and restoration of ecological balance. However, remediation can be destructive to the ecosystem [4], if the application is not properly addressed to meet the eco-friendly standard required to combat the contemporary issues of pollution [4, 5]. Most small-scale applications of bioremediation approaches using bioremediation agents such as bacteria, fungi, plants, and organic materials have been successful with variation in results, although bioremediation on a large scale has not been widely validated [4]. This chapter aims to propose a cost-effective and eco-friendly bioremediation strategies that could reduce or remove contaminants from the environment and thus stabilizing the ecosystem from heavy metal pollution and oil spills.

1.1 Bioremediation technologies

1.1.1 Bioattenuation or natural attenuation

There are a lot of different physical, chemical, and biological processes commonly termed bioattenuation, which make pollutants smaller in terms of their size and toxicity as well as how much of them there are. Some examples of these processes are sorption, volatilization, chemical or biological stabilization, and the transformation of contaminants. This entails removing pollutant concentrations from the surrounding through biological methods or perhaps incorporating (oxic and anoxic biodegradation, plant and animal sorption), physical occurrences (changes in weather conditions, dispersion, dilution, diffusion, volatilization, sorption/desorption), and chemical reactions (ion exchange, complexation, abiotic change) [6–8]. For instance, natural biodegradation and biotransformation are incorporated within the broader notion of common restriction [9, 10]. At the point when the site is contaminated with chemicals, the environment acts in 4 different approaches to facilitate remediation [11]:

- Microbes or microorganisms living in soil and groundwater may consume just a small number of chemical or manmade chemicals available as dietary nutrients. When they have completely digested the chemical, they can convert it to water and non-toxic gases.
- Chemical compounds can stick to or sorb to the soil, which prevents them from contaminating groundwater or escaping the location.
- As contamination travels through soil and groundwater, it can blend in with clean water. This diminishes or weakens the contamination.

- Certain chemicals, such as oil and solvents, can disappear, hence, they can transform from liquids to gases within the soil or groundwater. As a result, if these gases reach the earth surface via the air, they may be pulverized by sunlight.

Additionally, if natural attenuation is insufficiently rapid or complete, bioremediation will be accelerated or augmented via biostimulation, bioaugmentation, bioventing, or biopile [11, 12].

1.1.2 Biostimulation

This bioremediation approach invigorates the activity of native microbes by adjusting the environmental parameters or the introduction of nutrients [11, 13]. This is carried out with the incentive of natural or normally prevailing parasites or microbial communities [7, 11, 13]. Successive steps involve providing manures, development enhancements and minor elements. Also, by giving other natural prerequisites including pH, temperature and oxygen to enhance their digestion rate and degradation pathway [10, 12]. Similarly, the presence of pollutants even in small quantities can act as a stimulant by spinning for bioremediation proteins. Typically, this type of deterioration is followed by the provision of organic or inorganic nutrients and oxygen to promote the metabolism of native microbes for effective remediation [6]. These nutrients are the fundamental building blocks of life, enabling microorganisms to synthesize vital components such as enzymes, energy, and cell biomass required to degrade the toxin [6, 14]. However, nitrogen, phosphorous and carbon are significantly required to enhance metabolism.

1.1.3 Bioaugmentation

This procedure entails sequentially adding contaminant-degrading microbes (inherent/non-native/genetically modified) to improve the biodegradative efficiency of the native microbial community in the polluted site [8, 11]. Thus, to rapidly grow the natural microbial population and accelerate breakdown at the pollutant's location. Microorganisms that predominate in polluted sites on a global scale, may surely change significant amounts of harmful substances into non-poisonous structures. This process converts pollutants to by-products like carbon (IV) oxide and water, as well as metabolic intermediates that serve as critical nutrients for cell development [15, 16]. Microorganisms can also be isolated from the remediation environment, cultured autonomously, genetically engineered, and then reintroduced to the site [8, 11]. For persuade, all basic microbes are prevalent in locales where soil and groundwater are polluted with chlorinated ethenes, for example, tetrachloroethylene and trichloroethylene [7, 8, 11]. These are employed to facilitate the effective removal and conversion of these pollutants to non-poisonous ethylene and chloride by in situ microbes [10].

Additionally, genetically modified microbes have been shown to degrade a broad range of environmental contaminants effectively. Since the metabolic pathway can be altered to produce less puzzling and harmless end products [8, 17]. Genetically engineered microorganisms (GEM) have shown viability in bioremediation of soil, groundwater and activated sludge, proving effective degradation abilities of extensive integration of chemical and physical contaminations. GEMs have better enzyme abilities, which makes them better at breaking down a wide spectrum of aromatic hydrocarbons and making the soil more fertile [14, 18]. There are several types of hydrocarbon-degrading microorganisms that include the genera *Alanivorax* and

Bacillus, *Pseudomonas* and *Bravibacillus*, *Acinetobacter* and *Methylobacterium*, and *Candidauts* as well. Biodegradation of the benzene, toluene, ethylbenzene, and xylene (BTEX) isomers may be discovered in situ using the polymerase chain reaction, and nucleotide sequence analysis of BTEX degraders in the environment [7–9, 14, 18].

1.1.4 Bioventing

It is the practice of venting oxygen through the soil to encourage the development of natural or injected microbes and fungus in the soil by supplying oxygen to the soil microbes, which has been termed as bioventing [8, 11, 14]. The use of low air flowrates to supply sufficient oxygen to sustain microbial movement has long been a typical practice in aerobic degradation of substances, and it has been for many years. For example, several scientists have demonstrated that bioremediation of oil-contaminated soil utilizing bioventing may be achieved with reasonable success [19]. Consequently, petroleum residuals and their by-products are biodegraded, and volatile organic compounds, when destroyed, release vapors that slowly permeate through the biologically dynamic soil environment.

1.1.5 Biopiles

Biopile, also known as biocells, bioheaps, biomounds and composts piles are employed to minimize the toxicity of total petroleum hydrocarbon constituents via microbial respiration. Biopiles are an ex-situ bioremediation technology that consists of piling polluted soil onto a compost pile (biopiles) or cells (biocells) or mounds (biomounds) or heaps (bioheap) and stimulating oxic metabolism in the soil via aeration or introduction of minerals or nutrients, bulking agents, and subsequently confining it in a treatment bed with polyethylene material to avoid evaporation, surface runoff, and volatile emissions. Biopiles treatments can transform pollutants into low-toxic by-products through biological processes by utilizing already existing microorganisms to breakdown fuels and oils into carbon dioxide and water.

The biopile technology is made up of commercial roll-off dumpsters or containers that have been turned into fully contained bioremediation units. The biopile units have an impermeable liner to decrease the possibility of leachate movement to the subsurface ecosystem. Excavated soils are combined with soil additives and placed on a treatment area with leachate collecting devices and some type of aeration to maximize and regulate the rate of biodegradation. Air is introduced to the biopile mechanism of piping and pumps, which either power air into the heap under a specific tension or draw air through the heap under a negative tension [8, 20]. Microbial movement, for instance, can boost the adsorption and degradability of petroleum pollutants during funneling and siphoning operations. Biopiles, such as biocells, bioheaps, biomounds, and compost, might alleviate public concern about excavated soil contaminated by vigorously remediable hydrocarbons [8, 13, 19].

1.1.6 Phytoremediation

Utilizing plants for bioremediation is highly dependent on their ability to break down certain pollutants [21–24]. Phytoremediation is the process of utilizing plants to degrade, eliminate, or convert contaminants to less hazardous chemicals [25]. Even though plants have been used for soil purification for centuries, scientists have contributed to its advancement and expanded its scope of application throughout the years [7, 11, 13, 14, 17, 18]. This involves the removal of metals, pesticides, solvents, explosives, and raw petroleum, as well as a variety of other pollutants from soils, water (surface and subsurface), and vaporous contaminants [7, 11, 14]. When the plants have accumulated enough toxins, they are harvested and disposed of. **Figure 1** shows a graphical presentation of different types of phytoremediation as each mechanism is explained as follows:

i. Phytostabilization: this entails using plants to minimize soil erosion, so immobilizing contaminants by limiting their movement and accessibility in the soil via the plant roots. Additionally, it prevents metals from moving to the soil or the surface of underground water.

ii. Phytovolatilization: this involves the use of plants to minimize soil erosion, so immobilizing contaminants by limiting their movement and accessibility in the soil via the plant roots. Additionally, it prevents metals from moving to the soil or the surface of underground water.

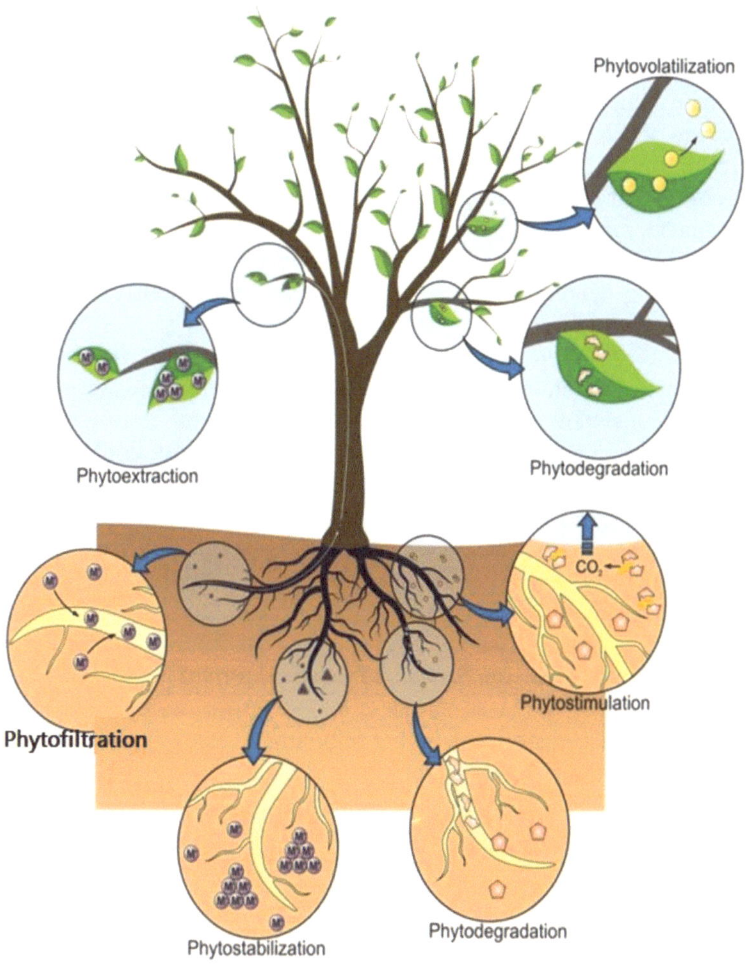

Figure 1.
Schematic presentation of phytoremediation process (adapted from [7, 11]).

iii. Phytodegradation: this process includes the degradation or modification of pollutants in the plant tissue by enzymes.

iv. Phytoextraction: this approach involved the extraction of contaminants from the soil and their accumulation in the shoots. Upon that, these plants' leaves are gathered, burned for energy, and the metals retrieved from the ash are regenerated.

v. Phytofiltration or rhizofiltration use roots to accumulate and sequester contaminants from polluted water.

vi. Phytostimulation or rhizodegradation: plant roots are employed to digest organic pollutants in the rhizosphere environment and through microbial activity.

1.1.7 Combinative bioremediation

This is when two or more bioremediation methods work together to remove contaminants from the environment. This kind of bioremediation technique can be effectively applied in a multi-contaminated environment. The combinative strategy most likely to be suitable and effective in boosting bioremediation of bauxite residue is a combination of bioaugmentation (incorporation of inocula) [8, 11] and biostimulation (introduction of nutrients to enhance the activity of microorganisms) of the indigenous community in bauxite residue [11, 13].

In this scenario, for instance, biostimulation using organic or inorganic compounds can be applied as the first or basic treatment while bioventing or bioaugmentation using engineered microbes can be applied subsequently as a secondary or tertiary treatment to facilitate the removal or degradation of recalcitrant compounds. Combinations of bioaugmentation and biostimulation have also proven effective, albeit they do not always show significant improvements over bioaugmentation alone. Given the nearly consistent advancement seen with bioaugmentation technology, it is anticipated that bioaugmentation will improve on the outcomes obtained so far with biostimulation for bauxite waste cleanup (provided an appropriate choice of the microbes and adequate trials are prioritized). Based on the simplicity of obtaining and introducing the inoculum, the most suited approach for future research and field trials is combinative bioremediation using biostimulation and bioaugmentation technology.

1.2 Bioremediation mechanisms for contaminant removal

Several bioremediation mechanisms for reducing or oxidizing contaminants have been discovered over time, such as adsorption, physio-biochemical (biosorption and bioaccumulation) bioleaching, biotransformation, biomineralization, and molecular mechanisms [7, 11].

1.2.1 Adsorption bioremediation mechanism

Environmental pollutants (both organic and inorganic) can be absorbed by microorganisms at specific sites in their cell structure that do not require the dissipation of energy. There are many various kinds of chemicals connected with bacterial cell walls, but their extracellular polymeric substances (EPS) are of particular importance since they have been shown to have significant effects on corrosive base characteristics and metal adsorption [10, 26]. Several studies on the metal binding behavior of EPS have revealed that it has a remarkable capacity to absorb complex metals by a variety of processes that combine ion exchange and micro-precipitation of metals [10, 13]. Bioremediation research and application are still limited in the present scenario due to a lack of understanding of the genetic traits and genome-level properties of the organisms used in metal adsorption, the metabolic route, and the kinetics of metal adsorption [7].

1.2.2 Physio-biochemical mechanism

In microscopic organisms, inhibition is advanced through two mechanisms: detoxifying (changing the detrimental metal's state and rendering it inaccessible) and dynamic efflux (siphoning poisonous heavy metals from cells) [7, 9]. In wastewater or soil, the fundamental redox (oxidation and reduction) reaction occurs between hazardous metals and microorganisms. Additionally, microbes oxidize heavy metals, causing them to lose electrons, which are recognized by active electron acceptors (nitrate, sulphate and ferric oxides) [26]. Additionally, the biosorption process, which consists of a biosorbent's increased affinity for sorbate (metal ions), is repeated till a balance between the two components is established [18, 26]. For instance, *Saccharomyces cerevisiaeacts* as a biosorbent for Zn (II) and Cd (II) removal via the ion exchange process [10, 26]. *Cunninghamella elegans* is also reported as a potential sorbent against substantial metals delivered by textile wastewater [12, 17].

Bioaccumulation is a term referring to the combination of active and passive techniques of hazardous metal bioremediation. Additionally, bioremediation may entail aerobic or anaerobic microbial activity [10, 12, 13]. Aerobic degradation frequently involves the addition of oxygen atoms to the reactions via monooxygenases, dioxygenases, hydroxylases, oxidative dehalogenases, or chemically active oxygen molecules produced via catalysts including ligninases or peroxidases [10–13]. Anaerobic contaminant corruptions comprise initial enactment reactions followed by oxidative degradation with the assistance of anaerobic electron acceptors. The act of Immobilization refers to the process of reducing the activation of significant metals in a polluted environment by modifying their physical or synthetic state [7, 12]. Microbes muster metals from polluted sites through leaching, filtering, chelation, methylation and redox transformation of harmful metals [12, 17]. Since significant metals cannot be entirely eliminated, the cycle modifies their oxidation state or organic complex to make them more soluble, less poisonous and precipitated [9, 14].

1.2.3 Bioleaching

In bioleaching, naturally occurring microorganisms such as bacteria and fungi solubilize metal sulphides and oxides from ores and secondary wastes. Adsorption, ion exchange, membrane separation, and selective precipitation are some of the processes used to purify solubilized metals. It is a cost-effective and environmentally beneficial technique because it consumes less energy and produces no hazardous gases. It has been applied to leach metals from low-grade ores, and it now provides a substantial global business in the extraction of metals like copper, cobalt, gold, nickel, uranium, zinc, and other elements [27].

1.2.4 Biotransformation

This is the procedure for altering the structure of a chemical substance to produce a molecule with higher polarity. Moreover, this metal-microbe interaction process converts hazardous metal and organic chemicals into a less poisonous form. This mechanism has emerged in microorganisms to assist them in adjusting to variations in their surroundings. Bacterial cells have a significant surface-volume ratio, a rapid pace of proliferation, a rapid rate of metabolic activities, and are easy to keep sterile [27]. As a result, they are perfect for biotransformation. Various methods, such as condensing and hydrolyses, forming new carbon bonds, isomerization, inserting functional groups, and oxidation, reduction, and methylation, can be used to attain this objective. Metals may be volatized, reducing their lethal nature, as a result of these interactions.

1.2.5 Biomineralization

Biomineralization refers to the mechanisms by which microbes produce minerals, and it can lead to metal extraction from solution, which can be used for decontamination and biorecovery. Dead biota and related products may also serve as a model for mineral deposition, with physicochemical parameters determining whether the process is reversible or not. There are several prevalent microbe-precipitated biominerals with unique chemical features such as high metal sorption capacities and redox catalysis. However, some biominerals can be deposited at nanoscale dimensions, resulting in additional physical, chemical, and biological features that can be used in practical applications [28].

1.2.6 Molecular mechanisms involved in bioremediation process

Different components of genetically altered bacteria, such as Deinococcus geothemalis, are active in the removal of heavy metals [9, 14, 18]. Hg^{2+} reduction has been recorded at high temperatures as a result of the expression of meroperon from *Escherichia coli* coded for Hg^{2+} reduction [18]. Two distinct components for Hg reduction by microscopic organisms ((*Klebsiella pneumonia* M426) are mercury volatilization by the decrease of Hg (II) to Hg (0) and mercury precipitation as insoluble Hg attributed to unstable thiol (H_2S) [7, 18]. Genetic of *Deinococcus radiodurans* (radiation tolerant bacterium) which usually decreases Cr (IV) to Cr (III) has been done for toluene (fuel hydrocarbon) reduction by cloned genes of *tod* and *xyl* operons of *Pseudomonas putida* [9, 11, 18]. Microbial metabolites including metal-bound coenzymes and siderophores are usually part of the degradation pathway.

2. Bioturbation

The promising bioremediation technique involves the application of bioturbators. Bioturbation is made up of a series of processes triggered by microbenthic fauna that influences sediment physicochemical characteristics and affects the microbial population which partake in the distribution of nutrients [29]. Bioturbation involves a series of activities such as the reworking of particles, bioirrigation, and other benthic biota related behaviors (i.e. nutrition mode and grazing by animals and organisms) that were responsible for transportation and distribution of porewater and particles along the water-sediments interface [30]. The distribution of dissolved contaminants can be a reworking of sediments by bioturbators through facilitating transportation and biomixing efficiency from overlying water and porewater to deep layers of the sediment [31–33].

The term "bioturbation" relates to the procedure of completely transforming dangerous hazardous substances into harmless or naturally occurring chemicals. Bioturbation can be done in situ (for example, in field conditions) or ex-situ (for instance, in a microcosm or under controlled conditions). Both scenarios entail the utilization of plants, parasites/fungi, and microorganisms as bioremediators for the biodegradation of toxic pollutants, even though individualized end product may be a different component [34–36]. Thus, complete breakdown of the contaminants by the bioremediators directly or indirectly may influence the residue structure [34, 37]. **Figure 2** presents significant types of contaminant improvement approaches by bioturbators (benthic fauna) in the contaminated environment to facilitate residue treatment.

Figure 2 illustrates the following:

i. Biodiffusors: this is performed through microorganisms' activities, which often result in the biomixing of uniform and irregular sediments over short separations, resulting in particle interchangeability via molecular diffusion.

ii. Upward conveyors: these are organisms that live vertically head-down in the sediments. They transfer particles from the residue's deep horizons to its surface. Gravity then returns the particles to the base under the influence of feces pellet agglomeration at the sediment surface.

iii. Downward conveyors: these are head-up feeders that actively pick and consume particles near the surface, as well as discharge in deeper residual layers.

iv. Regenerators: these microorganisms dive into the leftovers and constantly maintain burrows, so transferring dirt from depth to the surface.

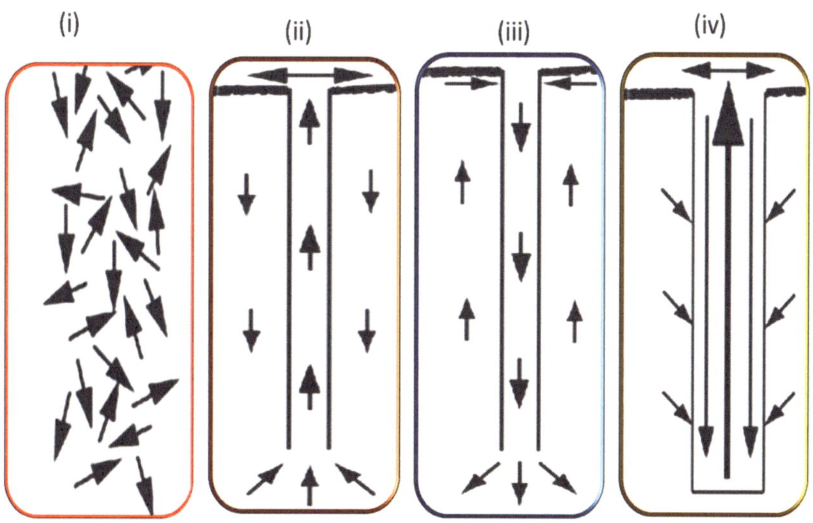

Figure 2.
Schematic representation of bioturbators activities in sediments (i) biodiffusors, (ii) upward conveyors, (iii) downward conveyors, and (iv) regenerators.

2.1 The role of bioturbation in bioremediation of organic and inorganic contaminants

The role and effectiveness of bioturbators in bioremediation is dependent on several conditions, such as the chemical type and quantities of contaminants, the physicochemical properties of the environment, and their accessibility to microbes [38]. Bioturbators are responsible for vital changes in the biological and physicochemical aspects of soils and water [38, 39]. Additionally, aerobic bioturbation can increase benthic digestion and supplement components by stimulating oxygen-consuming bacterial networks that are concerned with pollutant mitigation [8, 11]. In other words, bioturbators are well-suited for a dual-purpose mechanism, namely the production of degradative enzymes for specific contaminants and resistance or protection from significant relative dangerous substances such as heavy metals [15, 38, 39]. Controlling and simplifying bioremediation procedures is a difficult process due to a large number of components including the presence of a microbial community with the ability to detoxify pollutants, the contaminants' accessibility to the microbial community, and abiotic conditions (soil type, temperature, pH, oxygen or other electron acceptors, and substrates) [6, 16, 39].

Bioturbation influences the sediment-water interface's biological, physical, and chemical properties which accounts for the high rate of mineralization of organic matter in the aquatic environment [40]. This operation changes the sediment column distribution of the contaminants [41]. Bioturbation and biotransport can affect the physicochemical characteristics of sediments and sediment pollutants [42–44]. Bioturbation controls the organic matter and nutrient digestion enhances pollutant mobility and transformation [45–48]. The biosorption of organic contaminants into the organic matter during bioremediation reduces its bioavailability for plants (phytoremediation) or degrading organisms (bioaugmentation) [49]. Atrazine removal from sediments is promoted and positively influenced by the adjustment of organic matter and earthworm bioturbation activities, which increases contaminant bioavailability and atrazine sorption rate on their microsites [46, 50]. Previous studies reported positive contributions of earthworm bioturbation to organic pollutant transformation and biodegradation [51, 52] by modifying pore size and metabolism of degrading bacteria groups or accelerating mineralization in bioaugmented soils [50].

Moreover, several studies showed that bioturbation alters the physicochemical characteristics of the water-sediment boundary which promotes the bioavailability of inorganic pollutants to degrading organisms. This is achieved through the modification of sediment particle sizes, pore spaces, moisture content, nutrient content, turbidity, and total organic carbon of the vadose water-sediment [41, 43, 53]. Also, the bioturbation of benthic invertebrates through the mixing of sediments in the underground zone enhanced the electron acceptors (oxygen, nitrate and sulphate) entrance into the vadose zone which triggers geochemical changes that influence metal behavior [54]. The presence of these electron acceptors in the unsaturated zone can activate the RedOx reaction to change the chelating of metals affinities between liquid and solid phases to enhance the quantitative distribution and bioavailability of metal in the sediment [55]. The changes created by the bioturbation-attributed redox potentials, pH, organic content, pore spaces can affect metal sorption capacity and improve metal conversion from one phase to another e.g. Cd, Zn [56–58].

2.2 Factors that influence bioremediators or bioturbators for pollutant remediation

The activities of bioturbators are affected by some factors which modulate the rate of bioturbation for effective remediation of polluted environments. These factors include the variation in salinity, temperature, density, sediment grain size pH, and concentration.

2.2.1 Variation in salinity level

Variation in salinities in the aquatic environment can influence the metabolism of nutrient and metal releases [59, 60], whether naturally and/or through human-related activities. Remaili et al. [61] noted that hypersalinity has a negative effect on the larger bioturbators which affects the activities of benthic organisms. Gonzalez et al. [62] study found that the salinity levels and tolerance of various bioturbators are distinct. The findings however suggest that ammonia release in the aquatic environment is significantly modified due to the effect of modulating conditions and distinguished by a higher salinity than other nutrients such as phosphorus [62, 63].

2.2.2 Temperature variation

Regional variability in temperature is also a crucial factor that regulates the impact of bioturbation in pollutant remediation. In microbial response, metabolism, and degradation of organic matter and metals, temperatures played a fair modulatory function [64]. In the presence of bioturbation activities, the rise in temperatures increases the production of ammonium from the sediment, possibly due to the high level of hydrogenase in microbial species and the increased aerobic conditions in the sediment [64, 65]. Gonzalez et al. [62] reported that an increase in temperature is indirectly proportional to the nutrient dispersion as high temperature decreases nutrient flux (phosphorus) in the sediment but extreme temperatures may be devastating to the microbes. However, an increase in temperature corresponds to the increased rate of metal resuspension and metal solubility as a result of higher bioturbation rates [66, 67].

2.2.3 Bioturbator density

The bioturbator density influences bioturbation, control bioturbation efficiency for contaminant remediation, which correlates with the increased aerobic microbial activity and emission of pollutants. The increased bioturbation density increased phosphorus release and induced aerobic microbial activity but did not increase the release of ammonia. Animal density is a highly imperative factor, as study reveals that higher densities contribute toward greater degradation and mineralization of organic matter but may also increase nutrients in the overlying water and can, depending on the ecosystem studied, have counterproductive effects on recovery [66]. In response to pollution, the population of certain benthic species such as polychaetes [68] may increase as several systems are deprived of the use of other larger bioturbators.

2.2.4 Sediment grain size

Another element that influences the high level of organic matter and metals accumulation and the structure and metabolism of microbial communities and their metabolism is the sediment grain size [69, 70]. A recent study also shows a positive association between ammonia, phosphorus release, and aerobic microbial activity for the sediment grain size as Martinez-Garcia et al. [70] noted that the grain size showed less effect at low organic enrichments, but instead, at higher enrichments, coarse sediments contain less organic matter and nutrients while metabolism rate is enhanced. The contaminant bioavailability assessment can be affected by the susceptibility, grain size and behavior of microbes used in bioassays or observed on the ground, and the interaction between various species and microbial populations in highly polluted sediments depauperated by larger invertebrates [1, 71].

2.2.5 Contaminant pH and concentration in the sediment

The concentration of organic or inorganic contaminants is another factor that regulates the activities of the benthic organisms [72] which tend to either reduce or hinder the activities of the benthic organism at a high concentration, beyond the tolerable limit, which can result in the death of these organisms at extreme condition due to toxicity [5]. Benthic organisms have varying tolerance limits for sediment contaminations and tend to possess special features or activities (such as bioaccumulation or biosorption) to enable them to adapt and function effectively in high pollutant concentrations. For metal remediation, abiotic factor-like pH which works closely with concentration may be a crucial modulating variable that determines the impact of bioturbation in the marine environment which can alter metal speciation and reactivity [66].

Therefore, sediment properties like particle size and concentration as well as contaminant shape (sulphides or organic carbon) can affect the bioavailability of the contaminant. Also, in most environments, temperature and type of organism activity or population density can increase or decrease contaminant exposure or bioavailability for bioremediation [61, 73–75].

3. Challenges in bioremediation application

Notwithstanding the benefits (such as environmental friendliness, selectivity, adaptability, self-reproducibility, and the ability to recycle bioproducts) of the bioremediation technique, some setbacks have hindered the successful application of this technology. The delay of the operations and the complexity in managing the procedures are the two most significant disadvantages of this technique of treatment. Since the elimination of significant concentrations of heavy metals is a priority, and that the world has become more aware of the environmental concerns caused by other approaches, microbial procedures offer the most rational and long-term answer for treatment. As previously stated, while a variety of microbial contaminant bioremediation techniques to address contamination have been developed, their extensive use and application on a commercial scale are still restricted by some factors. A further point to mention is that the long-term viability of microbial decontamination is still a subject of significant importance, given the paucity of investigations into its long-term performance. Due to the extremely high accumulation of inorganic contaminants (heavy metals) in heavily inhabited places of the world, updating existing microbial bioremediation technologies to an industrial level by making the procedures quicker, more reusable, and easier to regulate will be a big issue in the future. Furthermore, another limitation of bioremediation is that not all substances are biodegradable while some hydrocarbon components are recalcitrant to microbial breakdown, which restricts the scope of the remediation technique. Even when a material is biodegradable, its downstream operation and breakdown can result in the production of harmful by-products in some situations.

4. Future approach of bioremediation technology

4.1 Genetically engineered microorganisms (GEMs)

The potential for microorganisms to remediate water and soil pollutants to increase treated water consumption and soil fertility for agricultural output is gaining attention [11, 38]. Recently, research has been conducted to enhance the application of altered organisms delineated specifically to boost their affectability toward hazardous metals [11, 16, 38]. An organism whose genetics have been transformed by the use of synthetic methods, which are driven by an artificial genetic exchange between bacteria, is referred to as a "genetically engineered microorganism [11, 18]. By developing GEM, genetic engineering has enhanced the application and disposal of hazardous wastes in laboratory settings. In addition, the following protocols must be considered during the GEM process: (a) alteration of enzyme selectivity and affinity, (b) pathway development and modulation, (c) bioprocess advancement, surveillance, and control, and (d) bioaffinity bioreporter sensor utilization for chemical detecting, toxicity reduction, and endpoint evaluation [13, 18].

As there are several possibilities for improving degradation performance through genetic engineering approaches, such as genetically controlling the rate kinetics of known metabolic pathways to increase degradation rate, or completely infusing bacterial strains with new metabolic pathways for the degradation of previously recalcitrant compounds [6, 8]. Despite important genes for microorganisms are carried on a single chromosome, defining the specific genes needed for the catabolism of some of these novel substrates may be carried on plasmids [18, 76, 77]. Plasmids were entangled in the catabolism process. As a result, GEM can be successfully used for biodegradation purposes, necessitating immediate research and large-scale deployment. Genetically engineer microbes offer the benefit of developing microbial strains which can tolerate unfriendly upsetting circumstances and can be utilized as a bioremediation tool under different and complicated natural conditions [18, 37, 76, 77]. Additionally, GEM has encouraged the development of "microbial biosensors" capable of precisely quantifying the degree of pollution in a contaminated site.

4.2 Engineered plants approach

The current advancement in omics technologies, including genomics, proteomics, transcriptomics, and metabolomics, play a critical role in finding characteristics that optimize remediation solutions [7, 11, 78]. Consequently, phytoremediation was developed, a process for eliminating toxins or their metabolites from plant tissues. This usually shortens the life of the plant and finally volatilizes the toxins into the atmosphere [78]. This disadvantage can be mitigated by managing plants' metal resistance, accumulation, and breakdown capacity in the presence of various inorganic toxins. To improve metal decomposition in plants, bacterial genes responsible for metal reduction can be integrated into plant tissues. As a result, plant-based bioremediation for a variety of significant metal poisons is cutting-edge due to its eco-friendliness. They are more effective at reducing dangerous substances than Physicochemical approaches, which are less environmentally friendly and potentially detrimental to human health [7, 8].

Notwithstanding, microbial genes can bridle in the transgenic plant for decontamination and collection of inorganic pollutants [7, 11]. The metal-detoxifying chelators, for example, metallothioneins and phytochelatins can give resistance to the plant by upgrading take-up, transport and amassing of different heavy metals [14, 78]. Similarly, transgenic plants with bacterial reductase can augment the volatilization of Hg and Se while absorbing the arsenic in plant shoots [17, 78]. Also, high-biomass-producing plants including poplar, willow and Jatropha can be applied for both phytoremediation and energy generation [7, 14, 26, 78]. Nonetheless, metals can only be removed from soil or water, which is why consuming metal-contaminated plants is advantageous. Thus, metal-accumulating biomasses should be properly preserved or disposed of to avoid posing an environmental hazard [20, 78].

4.3 Engineered *Rhizosphere* approach

Bioremediation methods include the introduction of growth stimulators (electron acceptors/donors) or nutrients to the rhizosphere to promote microbial growth and bioremediation characteristics of microbes or genetically engineered plants [6, 26, 78]. Multiple small organisms were generated with heavy metals by drainage using synthesized catalysts such as chromate and uranyl reductase in a particular rhizosphere [19, 26, 78]. Although genomics has been studied and applied mostly in microbial genetics and agriculture, such as genetic crops, and now serve as a bioremediation instrument [26, 76]. The application of genomics to bioremediation enables the microorganism to be dissected based on biochemical constraints as well as sub-atomic levels associated with the component [26, 76, 77].

4.4 Integrated bioturbation: phytoremediation process

Bioturbation is a very prolific and appealing technology for remediation, cleaning, management, and recovery of environmental contamination caused by microbial activity [11]. Furthermore, phytoremediation is successful at removing both inorganic and organic pollutants from residues or soils [7, 11, 12]. Nonetheless, investigation of resourceful bioremediation approaches for damaged aquatic environments that are based on these two processes to improve wastewater and soil treatment is necessary [10, 17]. Nonetheless, investigation of resourceful bioremediation technologies based on these two processes is important to improve soil and wastewater treatment [11, 17]. In addition, phytoremediation has been generally illustrated as a bioremediation process for heavy metals, such as lead, cadmium, copper, arsenic removal from contaminated soil or water [76, 77]. In essence, aquatic bioturbation combined with phytoremediation is a more effective and alternative method of removing heavy metals by improving cadmium transfers from overlying water to sediment and then into the root system of plants [15, 38].

Additionally, studies have demonstrated that earthworm movement greatly boosted phytoavailability by increasing soil macroporosity and generating cast around plant roots (**Figure 3**), implying that the physical effect of the earthworm's bioturbation is a viable mechanism [20, 26]. Interaction between plants and soil-dwelling microorganisms can also enhance phytoremediation known as rhizosphere bioremediation. The study by Leveque et al. [52] to investigate the contribution of

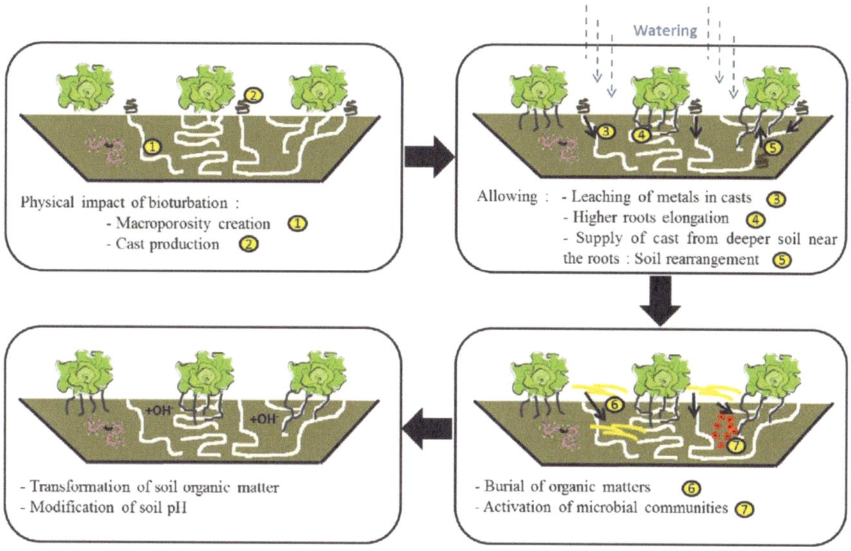

Figure 3.
Proffered approach to illustrate metal phytoavailability in earthworms' activities (adapted from [20, 26]).

earthworm (as bioremediator or bioturbation agent) to phytoremediation showed that earthworms significantly increased the phyto-availability of metal by generating soil macroporosity and developing cast near plant roots in which the main mechanism appears to be the physical impact of earthworm bioturbation. Moore et al. [21], demonstrated the contribution and the effect of bioturbators in the remediation of organic contaminants using the phytoremediation technique. In the study, *Typha latifolia* plant species recorded rapid growth in high pollutant concentrations in the environs due to its appreciable efficiency in the phytoaccumulation of contaminants from the sediments, which showed the ability to extract atrazine molecules by producing flux between the soil and the plant root. This plant was able to transform contaminants from atrazine to lower metabolites such as hydroxyatrazine, DEA and DIA [79].

4.5 Nano-biotechnology for bioremediation

The use of nanomaterials is extensively gaining attention for components remediation of heavy metals and recovery of valuable via nanotechnology [8, 34]. Conversely, nanobioremediation, which employs nanoparticles to stimulate microbial activity to clear hazardous chemicals from groundwater and soil [14, 17]. Not only can this nanotechnology greatly cut the cost of cleaning contaminated regions, but it also significantly shortens the procedure's duration. Metal chelating polymers require damaging solvents for mixing and ultrafiltration for division, which can be avoided by inventing metal limiting substances that can be reclaimed by adjusting their pH, temperature, or form, among other parameters [13, 19, 20]. One of the materials is nanoscale modified biopolymers, produced by microorganisms' intrinsic and protein structure, and whose size can be adjusted at the subatomic level [13]. For instance, polymers and magnetosomes are fabricated proteins for the remediation of infections, *Deinococcus radiodurans*, a radioactive-safe form of life, can resist radiation well past the naturally prevailing levels [13, 34, 37]. This is mostly used in radioactive waste remediation activities financed by the USA Division of Energy (DOE) [34, 38]. This technology seems to be very promising to address the rising concerns about heavy metals and other emerging contaminants in the aquatic environment.

4.6 Ecological engineering

The technique entails using ecological and environmental engineering expertise to create and monitor a sustainable ecosystem or biological system that benefits both humans and the environment. **Table 1** and **Figure 4** illustrate how to apply ecological engineering in a way that is more beneficial to humanity while maintaining the natural balance. Nevertheless, the majority of these technologies are typically designed with the following objectives in mind: (i) conservation, (ii) ecosystem restoration,

Ecological-engineering approaches	Terrestrial examples	Aquatic examples
Using ecosystems to solve a pollution problem	Phytoremediation	Wastewater wetland
Imitating or copying ecosystems to reduce or solve a problem	Forest restoration	Replacement wetland
Recovering an ecosystem after significant disturbance	Mine land restoration	Lake restoration
Existing ecosystems are modified in an ecologically sound way	Selective timber harvest	Biomanipulation
Using ecosystems for benefit without destroying the ecological balance	Sustainable agroecosystems	Multi-species aquaculture

Table 1.
Application of ecological engineering approach for terrestrial and aquatic systems.

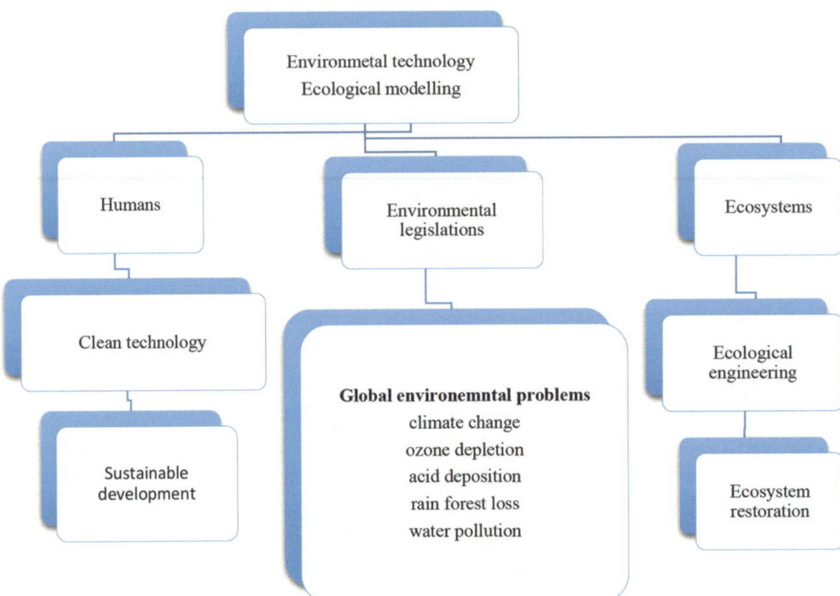

Figure 4.
Graphical representation of ecological engineering application to balance the ecosystem.

(iii) expanding ecological systems to the quantity, quality, and maintainability of their production, and (iv) assembling new ecological systems that would provide routine types of assistance [16, 39, 76, 77, 80].

5. Conclusion

Bioremediation is a cutting-edge and promising approach for treating contaminated soil and water. Microorganisms are also known to generate and use a variety of detoxification methods, including biosorption, bioaccumulation, biotransformation, and biomineralization for the remediation of the contaminated site during the bioremediation process. However, recent bioremediation research, such as bioturbation, which uses live organisms (macrofauna) directly or indirectly with the environment to eliminate toxins, is gaining momentum. The use of organisms to detoxify and recover polluted soil and water has emerged as the most robust, straightforward, and profitable technique. Microorganisms in water and soil have been studied and equipped to eliminate or detoxify harmful compounds discharged into the ecosystem due to anthropogenic processes such as mineral mining, oil and gas production, pesticides, pigments, plastic, organic solvents, fuel, and industrial operations. Nevertheless, a lack of data on microorganisms' cell reactivity to minor components and heavy metal poisons precludes their successful implementation. As such, the application of molecular genetic technology will enhance the efficiency and address most of the challenges in the large scale application of bioremediation technology.

Author details

Donald Tyoker Kukwa[1], Felicia Omolara Afolabi[1], Emmanuel Kweinor Tetteh[1], Ifeanyi Michael Smarte Anekwe[2]* and Maggie Chetty[1]

1 Department of Chemical Engineering, Durban University of Technology, Durban, South Africa

2 School of Chemical and Metallurgical Engineering, University of the Witwatersrand, Johannesburg, South Africa

*Address all correspondence to: anekwesmarte@gmail.com

References

[1] Chapman PM, Wang F. Assessing sediment contamination in estuaries. Environmental Toxicology and Chemistry. 2001;**20**(1):3-22

[2] Johnston EL, Roberts DA. Contaminants reduce the richness and evenness of marine communities: A review and Meta-analysis. Environmental Pollution. 2009;**157**(6):1745-1752

[3] Jain S, Arnepalli D. Biominerlisation as a remediation technique: A critical review. Geotechnical Characterisation and Geoenvironmental Engineering. 2019;**16**:155-162

[4] Boopathy R. Bioremediation of explosives contaminated soil. International Biodeterioration & Biodegradation. 2000;**46**(1):29-36

[5] Boopathy R. Factors limiting bioremediation technologies. Bioresource Technology. 2000;**74**(1):63-67

[6] Ron EZ, Rosenberg E. Biosurfactants and oil bioremediation. Current Opinion in Biotechnology. 2002;**13**(3):249-252. DOI: 10.1016/S0958-1669(02)00316-6

[7] Kumar A, Joshi V, Dhewa T, Bisht B. Review on bioremediation of polluted environment: A management tool. International Journal of Environmental Sciences. 2011;**1**(6):1079-1093

[8] Megharaj M, Ramakrishnan B, Venkateswarlu K, Sethunathan N, Naidu R. Bioremediation approaches for organic pollutants: A critical perspective. Environment International. 2011 Nov 1;**37**(8):1362-1375. DOI: 10.1016/j.envint.2011.06.003

[9] Watanabe K. Microorganisms relevant to bioremediation. Current Opinion in Biotechnology. 2001;**12**(3):237-241. DOI: 10.1016/S0958-1669(00)00205-6

[10] Viswanath B, Rajesh B, Janardhan A, Kumar AP, Narasimha G. Fungal laccases and their applications in bioremediation. Enzyme Research. 2014;**2014**:1-21. DOI: 10.1155/2014/163242

[11] Adams GO, Fufeyin PT, Okoro SE, Ehinomen I. Bioremediation, biostimulation and bioaugmention: A Review. International Journal of Environmental Bioremediation & Biodegradation. 2015;**3**(1):28-39. DOI: 10.12691/ijebb-3-1-5

[12] Iwamoto T, Nasu M. Current bioremediation practice and perspective. Journal of bioscience and bioengineering. 2001;**92**(1):1-8

[13] Dixit R, Wasiullah, Malaviya D, Pandiyan K, Singh UB, Sahu A, et al. Bioremediation of heavy metals from soil and aquatic environment: An overview of principles and criteria of fundamental processes. Sustainability (Switzerland). 2015;**7**(2):2189-2192. DOI: 10.3390/su7022189

[14] Dzionek A, Wojcieszyńska D, Guzik U. Natural carriers in bioremediation: A review. Electronic Journal of Biotechnology. 2016;**23**:28-36. DOI: 10.1016/j.ejbt.2016.07.003

[15] Vidali M. Bioremediation. An overview. Pure and Applied Chemistry. 2001;**73**(7):1163-1172. DOI: 10.1351/pac200173071163

[16] Mary Kensa V. Bioremediation—An overview. Journal of Industrial Pollution Control. 2011;**27**(2):161-168

[17] Ayangbenro AS, Babalola OO. A new strategy for heavy metal polluted

environments: a review of microbial biosorbents. International journal of environmental research and public health. 2017;**14**(1):94

[18] Pieper DH, Reineke W. Engineering bacteria for bioremediation. Current Opinion in Biotechnology. 2000;**11**(3):262-270. DOI: 10.1016/S0958-1669(00)00094-X

[19] Samanta SK, Singh OV, Jain RK. Polycyclic aromatic hydrocarbons: Environmental pollution and bioremediation. Trends in Biotechnology. 2002;**20**(6):243-248. DOI: 10.1016/S0167-7799(02)01943-1

[20] Farhadian M, Vachelard C, Duchez D, Larroche C. In situ bioremediation of monoaromatic pollutants in groundwater: A review. Bioresource Technology. 2008;**99**(13):5296-5308. DOI: 10.1016/j.biortech.2007.10.025

[21] Moore MT, Tyler HL, Locke MA. Aqueous pesticide mitigation efficiency of *Typha latifolia* (L.), *Leersia oryzoides* (L.) Sw., and *Sparganium americanum* Nutt. Chemosphere. 2013;**92**(10):1307-1313

[22] Fulekar M. Global status of environmental pollution and its remediation strategies. Bioremediation Technology. Dordrecht: Springer; 2010. pp. 1-6

[23] Murphy IJ, Coats JR. The capacity of switchgrass (*Panicum virgatum*) to degrade atrazine in a phytoremediation setting. Environmental Toxicology and Chemistry. 2011;**30**(3):715-722

[24] Qu M et al. Distribution of atrazine and its phytoremediation by submerged macrophytes in lake sediments. Chemosphere. 2017;**168**:1515-1522

[25] Vangronsveld J, Herzig R, Weyens N, Boulet J, Adriaensen K, Ruttens A, et al. Phytoremediation of contaminated soils and groundwater: Lessons from the field. Environmental Science and Pollution Research. 2009;**16**(7):765-794

[26] Malik A. Metal bioremediation through growing cells. Environment International. 2004;**30**(2):261-278. DOI: 10.1016/j.envint.2003.08.001

[27] Tayang A, Songachan L. Microbial bioremediation of heavy metals. Current Science. 2021;**120**(6):00113891

[28] Gadd GM, Pan X. Biomineralization, Bioremediation and Biorecovery of Toxic Metals and Radionuclides. Taylor & Francis; 2016;**33**(3-4):175-178

[29] Biles CL, Paterson DM, Ford RB, Solan M, Raffaelli DG. Bioturbation, ecosystem functioning and community structure. Hydrology and Earth System Sciences. 2002;**6**(6):999-1005

[30] Kristensen E, Penha-Lopes G, Delefosse M, Valdemarsen T, Quintana CO, Banta GT. What is bioturbation? The need for a precise definition for fauna in aquatic sciences. Marine Ecology Progress Series. 2012;**446**:285-302

[31] He Y, Men B, Yang X, Li Y, Xu H, Wang D. Relationship between heavy metals and dissolved organic matter released from sediment by bioturbation/bioirrigation. Journal of Environmental Sciences. 2019;**75**:216-223

[32] Teal LR, Parker ER, Solan M. Coupling bioturbation activity to metal (Fe and Mn) profiles in situ. Biogeosciences. 2013;**10**(4):2365-2378

[33] Timmermann K, Banta GT, Klinge L, Andersen O. Effects of bioturbation on

the fate of oil in coastal sandy sediments–An in situ experiment. Chemosphere. 2011;**82**(10):1358-1366

[34] Azubuike CC, Chikere CB, Okpokwasili GC. Bioremediation techniques–classification based on site of application: Principles, advantages, limitations and prospects. World Journal of Microbiology and Biotechnology. 2016;**32**(11):1-8. DOI: 10.1007/s11274-016-2137-x

[35] Concetta Tomei M, Daugulis AJ. Ex situ bioremediation of contaminated soils: An overview of conventional and innovative technologies. Critical Reviews in Environmental Science and Technology. 2013;**43**(20):2107-2139. DOI: 10.1080/10643389.2012.672056

[36] Guerin TF. Ex-situ bioremediation of chlorobenzenes in soil. Journal of Hazardous Materials. 2008;**154**(1-3): 9-20. DOI: 10.1016/j.jhazmat.2007.09.094

[37] Juwarkar AA, Singh SK, Mudhoo A. A comprehensive overview of elements in bioremediation. Reviews in Environmental Science and Biotechnology. 2010;**9**(3):215-288. DOI: 10.1007/s11157-010-9215-6

[38] Hazen TC. Bioremediation. In: The Microbiology of the Terrestrial Deep Subsurface. 1997;**1**:1-20. DOI: 10.1201/9781351074568

[39] Boopathy R. Factors limiting bioremediation technologies. Bioresource Technology. 2000;**74**(1):63-67. DOI: 10.1016/S0960-8524(99)00144-3

[40] Gerino M, Aller R, Lee C, Cochran J, Aller J, Green M, et al. Comparison of different tracers and methods used to quantify bioturbation during a spring bloom: 234-thorium, luminophores and chlorophylla. Estuarine, Coastal and Shelf Science. 1998;**46**(4):531-547

[41] Anschutz P, Ciutat A, Lecroart P, Gérino M, Boudou A. Effects of tubificid worm bioturbation on freshwater sediment biogeochemistry. Aquatic Geochemistry. 2012;**18**(6):475-497

[42] Gilbert F, Stora G, Cuny P. Functional response of an adapted subtidal macrobenthic community to an oil spill: Macrobenthic structure and bioturbation activity over time throughout an 18-month field experiment. Environmental Science and Pollution Research. 2015;**22**(20):15285-15293

[43] Mermillod-Blondin F, Rosenberg R, François-Carcaillet F, Norling K, Mauclaire L. Influence of bioturbation by three benthic infaunal species on microbial communities and biogeochemical processes in marine sediment. Aquatic Microbial Ecology. 2004;**36**(3):271-284

[44] Pigneret M, Mermillod-Blondin F, Volatier L, Romestaing C, Maire E, Adrien J, et al. Urban pollution of sediments: Impact on the physiology and burrowing activity of tubificid worms and consequences on biogeochemical processes. Science of the Total Environment. 2016;**568**:196-207

[45] Farenhorst A, Topp E, Bowman B, Tomlin A. Earthworm burrowing and feeding activity and the potential for atrazine transport by preferential flow. Soil Biology and Biochemistry. 2000;**32**(4):479-488

[46] Farenhorst A, Topp E, Bowman B, Tomlin A. Earthworms and the dissipation and distribution of atrazine in the soil profile. Soil Biology and Biochemistry. 2000b;**32**(1):23-33

[47] Hoelker F, Vanni MJ, Kuiper JJ, Meile C, Grossart H-P, Stief P, et al. Tube-dwelling invertebrates: Tiny

ecosystem engineers have large effects in lake ecosystems. Ecological Monographs. 2015;**85**(3):333-351

[48] McCall P, Fisher J. Effects of tubificid oligochaetes on physical and chemical properties of Lake Erie sediments. In: Aquatic Oligochaete Biology. Boston, MA: Springer; 1980. pp. 253-317

[49] Binet F, Kersanté A, Munier-Lamy C, Le Bayon R-C, Belgy M-J, Shipitalo MJ. Lumbricid macrofauna alter atrazine mineralization and sorption in a silt loam soil. Soil Biology and Biochemistry. 2006;**38**(6):1255-1263

[50] Monard C, Martin-Laurent F, Vecchiato C, Francez A-J, Vandenkoornhuyse P, Binet F. Combined effect of bioaugmentation and bioturbation on atrazine degradation in soil. Soil Biology and Biochemistry. 2008;**40**(9):2253-2259

[51] Kersanté A, Martin-Laurent F, Soulas G, Binet F. Interactions of earthworms with atrazine-degrading bacteria in an agricultural soil. FEMS Microbiology Ecology. 2006;**57**(2):192-205

[52] Leveque T, Capowiez Y, Schreck E, Xiong T, Foucault Y, Dumat C. Earthworm bioturbation influences the phytoavailability of metals released by particles in cultivated soils. Environmental Pollution. 2014;**191**:199-206

[53] Ciutat A, Widdows J, Readman JW. Influence of cockle Cerastoderma edule bioturbation and tidal-current cycles on resuspension of sediment and polycyclic aromatic hydrocarbons. Marine Ecology Progress Series. 2006;**328**:51-64

[54] Aller JY, Woodin SA, Aller RC. Organism-sediment Interactions. Columbia: University of South Carolina Press eSC SC; 2001

[55] De Jonge M, Teuchies J, Meire P, Blust R, Bervoets L. The impact of increased oxygen conditions on metal-contaminated sediments part I: Effects on redox status, sediment geochemistry and metal bioavailability. Water Research. 2012;**46**(7):2205-2214

[56] Cheng J, Wong MH. Effects of earthworms on Zn fractionation in soils. Biology and Fertility of Soils. 2002;**36**(1):72-78

[57] Yu C, Zhang J, Pang XP, Wang Q, Zhou YP, Guo ZG. Soil disturbance and disturbance intensity: Response of soil nutrient concentrations of alpine meadow to Plateau pika bioturbation in the Qinghai-Tibetan Plateau, China. Geoderma. 2017;**307**:98-106

[58] Zorn MI, Van Gestel CA, Eijsackers H. The effect of *Lumbricus rubellus* and *Lumbricus terrestris* on zinc distribution and availability in artificial soil columns. Biology and Fertility of Soils. 2005;**41**(3):212-215

[59] Nowicki BL. The effect of temperature, oxygen, salinity, and nutrient enrichment on estuarine denitrification rates measured with a modified nitrogen gas flux technique. Estuarine, Coastal and Shelf Science. 1994;**38**(2):137-156

[60] Rysgaard S, Thastum P, Dalsgaard T, Christensen PB, Sloth NP. Effects of salinity on NH 4+ adsorption capacity, nitrification, and denitrification in Danish estuarine sediments. Estuaries. 1999;**22**(1):21-30

[61] Remaili TM, Simpson SL, Bennett WW, King JJ, Mosley LM, Welsh DT, et al. Assisted natural recovery of hypersaline sediments: Salinity

thresholds for the establishment of a community of bioturbating organisms. Environmental Science: Processes & Impacts. 2018;**20**(9):1244-1253

[62] Gonzalez SV, Johnston E, Gribben PE, Dafforn K. The application of bioturbators for aquatic bioremediation: Review and meta-analysis. Environmental Pollution. 2019;**250**:426-436

[63] Magalhães CM, Joye SB, Moreira RM, Wiebe WJ, Bordalo AA. Effect of salinity and inorganic nitrogen concentrations on nitrification and denitrification rates in intertidal sediments and rocky biofilms of the Douro River estuary, Portugal. Water Research. 2005;**39**(9):1783-1794

[64] Bernard G, Duchene J-C, Romero-Ramirez A, Lecroart P, Maire O, Ciutat A, et al. Experimental assessment of the effects of temperature and food availability on particle mixing by the bivalve *Abra alba* using new image analysis techniques. PLoS One. 2016;**11**(4):e0154270

[65] Berkenbusch K, Rowden AA. Factors influencing sediment turnover by the burrowing ghost shrimp Callianassa filholi (Decapoda: Thalassinidea). Journal of Experimental Marine Biology and Ecology. 1999;**238**(2):283-292

[66] Amato ED, Simpson SL, Remaili TM, Spadaro DA, Jarolimek CV, Jolley DF. Assessing the effects of bioturbation on metal bioavailability in contaminated sediments by diffusive gradients in thin films (DGT). Environmental Science & Technology. 2016;**50**(6):3055-3064

[67] Andres S, Ribeyre F, Boudou A. Effects of temperature and exposure duration on transfer of cadmium between naturally contaminated sediments and burrowing mayfly nymphs (*Hexagenia rigida*). Archives of Environmental Contamination and Toxicology. 1998;**35**(2):295-301

[68] Dafforn KA, Kelaher BP, Simpson SL, Coleman MA, Hutchings PA, Clark GF, et al. Polychaete richness and abundance enhanced in anthropogenically modified estuaries despite high concentrations of toxic contaminants. PLoS One. 2013;**8**(9):e77018

[69] Jackson CR, Weeks AQ. Influence of particle size on bacterial community structure in aquatic sediments as revealed by 16S rRNA gene sequence analysis. Applied and Environmental Microbiology. 2008;**74**(16):5237-5240

[70] Martinez-Garcia E, Carlsson MS, Sanchez-Jerez P, Sánchez-Lizaso JL, Sanz-Lázaro C, Holmer M. Effect of sediment grain size and bioturbation on decomposition of organic matter from aquaculture. Biogeochemistry. 2015;**125**(1):133-148

[71] Eggleton J, Thomas KV. A review of factors affecting the release and bioavailability of contaminants during sediment disturbance events. Environment International. 2004;**30**(7):973-980

[72] Mulsow S, Landrum P, Robbins J. Biological mixing responses to sublethal concentrations of DDT in sediments by *Heteromastus filiformis* using a 137Cs marker layer technique. Marine Ecology Progress Series. 2002;**239**:181-191

[73] Cadée GC, Checa AG, Rodriguez-Tovar FJ. Burrows of Paragnathia (Crustacea: Isopoda) and Bledius (Arthropoda: Staphylinidae) enhance cliff erosion. 2001;**8**(3-4):255-260. DOI: 10.1080/10420940109380193

[74] Levinton J. Bioturbators as ecosystem engineers: Control of the sediment fabric, inter-individual interactions, and

material fluxes. In: Linking Species & Ecosystems. Boston, MA: Springer; 1995. pp. 29-36

[75] Remaili TM, Simpson SL, Amato ED, Spadaro DA, Jarolimek CV, Jolley DF. The impact of sediment bioturbation by secondary organisms on metal bioavailability, bioaccumulation and toxicity to target organisms in benthic bioassays: Implications for sediment quality assessment. Environmental Pollution. 2016;**208**:590-599

[76] Gadd GM. Metals, minerals and microbes: Geomicrobiology and bioremediation. Microbiology. 2010;**156**(3):609-643. DOI: 10.1099/mic.0.037143-0

[77] Lovley DR. Cleaning up with genomics: Applying molecular biology to bioremediation. Nature Reviews Microbiology. 2003;**1**(1):35-44. DOI: 10.1038/nrmicro731

[78] Chibuike GU, Obiora SC. Heavy metal polluted soils: Effect on plants and bioremediation methods. Applied and Environmental Soil Science. 2014;**2014**:1-12. DOI: 10.1155/2014/752708

[79] Mezzari MP, Schnoor JL. Metabolism and genetic engineering studies for herbicide phytoremediation. In: Phytoremediation Rhizoremediation. Dordrecht: Springer; 2006. pp. 169-178

[80] Singh R, Paul D, Jain RK. Biofilms: Implications in bioremediation. Trends in Microbiology. 2006;**14**(9):389-397. DOI: 10.1016/j.tim.2006.07.001

Chapter 3

Recent Applications of Bioremediation and Its Impact

Amara Dar and Arooj Naseer

Abstract

Socioeconomic concerns have increased the technology dependence to facilitate the increasing population on the earth. Number of anthropogenic sources are responsible for contaminating the natural environment. Effluents from industries contain many toxicants that cause lethal effects on human and animal life on earth. Many techniques are used so far for the abatement of such pollutants from the environment. As "nature heals itself" so dealing with such problems with bioremediation utilizing the invisible workers (microorganisms), plants and enzymes can help minimize and get rid of such pollutants. It is a greener way to conserve the environment and get rid of such awful substances. Bioremediation can help to get rid of contaminants either by in situ or ex situ approach. By using both ways, either ex situ or in situ, the decontamination of the environment can be successfully done. Using various plant materials and microorganisms by tailoring the surrounding environment to make it suitable for rectifying the contaminant issue is the main goal of bioremediation.

Keywords: bioremediation, microorganisms, enzymes, plants, sustainable development

1. Introduction

Bioremediation is a process of converting harmful substances to environmentally safe substances by the action of the invisible workforce. This invisible workforce is the number of microorganisms working in sequence to degrade environmentally toxic substances. Bioremediation works with detoxification and eradication of chemically diverse and physically hazardous materials that cause a threat to the natural existence of the environmental setup.

Ecologically discussing bioremediation refers to the interaction between three factors; Contaminant, invisible workforce, and environment, as shown in **Figure 1**. Interaction of these factors in turn ensures the mobility of contaminant in the environment, the presence of suitable conditions to degrade the contaminant, and eradication or degradation of the contaminant by converting it into an environmentally friendly substance. Mobility or bioavailability of any contaminant is about the ease with which the contaminant is available to microorganisms. The microorganisms need a suitable set of conditions (like; as availability of electron acceptors, pH, and availability of nutrients) to function well and convert the environmentally harmful substances to environmentally benign substances. The biodegradability of the contaminant depends

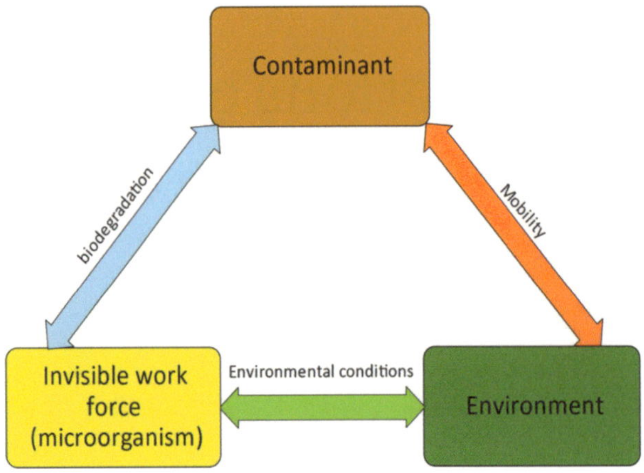

Figure 1.
Ecological interpretation of factors governing bioremediation [1].

upon the presence of suitable microorganisms to eradicate that contaminant under the required conditions [2].

Eradication of contaminants depends primarily on the nature of the contaminant, which may include pesticides, herbicides, heavy metals, hydrocarbons, sewage, plastics, etc. nature of the contaminant, degree of contamination, environmental factors, contaminated sites, cheap policies for conserving environment are important selection study criteria that are considered while choosing any bioremediation technique [3, 4]. Although it is important to properly plan the selection criteria but other factors that involve the aerobic and anaerobic nature of the area under study, pH, and moisture content are equally important to be considered. Bioremediation strategies make it possible to increase the efficacy of the contaminant removal process. These strategies may be ex situ or in situ. Mostly, the bioremediation techniques work for the removal of hydrocarbon contaminating species from soil or water [5–8]. Various other cost-effective techniques can be efficiently applied to the contamination sites for the removal of hydrocarbons [9].

2. Types of bioremediation

Majorly bioremediation can be categorized as:

2.1 Ex situ bioremediation

Ex situ bioremediation involves the excavation of contaminants and transporting them to the treatment sites above the ground. Indigenous microorganisms in the soil act as the remediating agents provided other environmental factors are kept monitored. This method can be tailored by changing the decay conditions and maintaining the optimum conditions required for microorganisms to work efficiently. In some conditions, the amendments are added to the soil. There are various types of ex situ techniques that include, biopiles, composting, windrow, landfarming, and slurry reactors [10].

*De*merits associated with ex situ bioremediation are its being expensive in terms of excavation and solid handling, fractionation and screening, and treatment till final disposal. The contaminant may be either solid or liquid. On the basis of phases of contaminant material, ex situ method can be a solid phase or slurry phase. In case of the solid phase method; waste in the form of solid-like agricultural waste or domestic, sewage sludge, industrial waste, and municipal solid waste are treated to get compost, which is further employed for the conditioning of soil. Treatment is done prior to compost formation to enhance the biological treatment potential. Various physico-chemical and biological factors of the site under study are considered for this purpose. Organic material thus presents or added to the soil act as the source of carbon for microorganisms. Depending on the availability of oxygen and suitable working pH the enzymes secreted by microorganisms detoxify the surrounding area. Applying ex situ treatment to a site that has some compositional limitations or nutrient deficiency for microbial activity needs tailoring of the site by adding site-specific compost. Adjusting pH and water availability to the bioremediating site ensures the efficiency of microbial colonies at ex situ operating sites. The slurry phase method is applicable to municipal wastewater. In this method supply of air, maintenance of proper pH, temperature, micronutrients are needed for the growth of microbial colonies [11].

2.2 In situ bioremediation

In situ bioremediation is the subsurface treatment of contaminants by the biological system of that area. These are considered sustainable methods as they do not require any excavation and transportation of contaminants. Some in situ bioremediation techniques like biosparging, phytoremediation, and bioventing, have been enhanced to get good outcomes for onsite decontamination while some other techniques like natural attenuation or intrinsic bioremediation proceed without any enhancement. In situ bioremediation techniques have been successfully used to treat chemically contaminated sites like; as industrial effluents dumping sites containing dyes, chlorinated solvents, hydrocarbons polluted sites, and heavy metals [12–14].

In situ bioremediation works with the abatement of contaminants by disrupting the minimum area and is a continuous and economical treatment method for soil and water. It can be intrinsic or engineered. Intrinsic bioremediation involves the conversion of contaminant to nontoxic form by the microbial communities naturally present in soil and water. The detoxifying potential of these microbial communities must be tested in laboratories so that outcomes can be configured accordingly. Various working conditions or requirements are there for intrinsic bioremediation to be fruitful. Annual water flow through the area understudy determines the presence of various minerals and pH of that soil which in turn tells about the working of microbes under such conditions. The presence of heavy metals hindered the growth of microorganisms present in the soil and water. The time of exposure of microorganisms to the contaminant is also an important parameter that should be studied at a pilot scale before conducting the bioremediation on a wide surface area. Although intrinsic bioremediation shows very promising decontamination results, but the limiting factor is when working conditions and environmental factors/site conditions do not favor microbial growth. In such cases, engineered bioremediation replaces intrinsic bioremediation. This type of bioremediation technique accelerates the growth of microbial colonies by providing suitable physicochemical growth conditions. The availability of oxygen, nutrients, and electron acceptors like sulfates and nitrates increase the onsite growth of microbes. I*n situ* bioremediation is laborious as compared to other methods. The outcomes of this method are highly environment-dependent. Continuous availability and replacement of nutrients must be ensured for efficient working genetically engineered microorganisms [11].

3. Factors influencing bioremediation

There are various factors that needed to be optimized to ensure the success of bioremediation processes [15].

3.1 Nature and concentration of contaminant

The concentration and nature of the contaminant are among one of the limiting factors for the bioremediation process. The presence of heavy metal impurities inhibits the growth of microbes and as a result of which it tends to inhibit the bioremediation process. Similarly, increased concentration of contaminant in the particular bioremediating site effect the microbial colonies both in terms of growth and enzymatic functioning of the microbes.

3.2 Nutrient availability

The presence of essential nutrients required for the growth and working of microbes is necessary for the proper outcomes of bioremediation. Nitrates, phosphates, and various electron transport sources are needed to be present in the soil or water environment where bioremediation is to be carried out.

3.3 Factors associated with contaminated site

3.3.1 pH

The pH of the soil or water determines the nature of the species present in it, as it can cause a change in the chemical composition due to the basic and acidic nature of the site. Generally, the pH ranges from 5 to 9 is considered optimum for the working of various microbial colonies. As biological reactions are all pH-sensitive so existence and functioning of enzymes are highly pH-sensitive.

3.3.2 Temperature

It is one of the important factors that defines the moisture content and chemical composition of the site. Generally, the temperature between 20 and 40° is considered good for the efficient performance of microbes.

3.3.3 O_2 availability and moisture content

The availability of oxygen is a very important factor also, as it ensures the oxidative and reducing environment in both soil and water environment. The nature of soil affects the aeration of the bioremediating site. Soil rich in sand and gravel content helps to retain moisture as well as aerate the soil well. More heavy the clay form or soil rich in organic content reduce the availability of oxygen and thereby inhibiting the functioning of microbes.

4. Various applications of bioremediation

Due to the increased population that is ultimately leading to pollution, anthropogenic activities have negative effects on ecosystems. Bioremediation is a purification technique to remove toxic waste from a polluted environment. Bioremediation is specifically helpful for decomposition, eradication, immobilization, or detoxification of variable chemical wastes and physical hazardous materials from the surrounding through the all-inclusive and action of microorganisms. The main principle is degrading and converting pollutants to less toxic forms. There are two approaches for bioremediation, in situ and ex situ. In situ methods involve treatment of the contaminated material at the site, whereas when the material is physically removed to be treated elsewhere it is referred to as ex situ. Bioremediation can occur naturally or stimulated, e.g. by the application of fertilizers (biostimulation), by the addition of similar microbe strains, the effectiveness of the resident microbe population to degrade contaminants may be increased. Every type of contaminant cannot be disposed of by means of microorganisms. Heavy metal contaminants, e.g. Cd^{2+} and Pb^{2+}, tend to resist interception by microorganisms. Bioremediation is the most effective, economical, eco-friendly management tool to manage the polluted environment. All bioremediation techniques have their own advantage and disadvantage because it has their own specific applications.

Bioremediation, an appropriate method, can be applied to different states of matter in the environment.

- Soils, sediment, and sludge as solids.
- Groundwater, surface water, and industrial wastewater as liquids.
- Industrial air emissions as gases.
- Saturated and vadose zones as sub-surface environments.

The biological community exploited for bioremediation generally consists of the native soil microflora. However, higher plants can also be manipulated to enhance toxicant removal (phytoremediation), especially for remediation of metal contaminated soils.

There are different types of bioremediants used for bioremediation. We can classify its applications on basis of its bioremediants.

4.1 Mycoremediation

It is an important form of bioremediation by the use of fungi. It is a cheaper method of remediation, and it does not usually require expensive equipment. Fungi are an excellent source to remove toxic pollutants from the environment and easily colonize both biotic and abiotic surfaces [16]. The most suitable fungi to be used in soil remediation are basidiomycetes and the ecological groups of saprotrophic and biotrophic fungi. Treu and Falandysz [17] various steps are involved in mycoremediation as shown in **Figure 2**.

- Fungi freely present in the soil, or in symbiotic association with plant roots (ectomycorrhizal and endomycorrhiza).

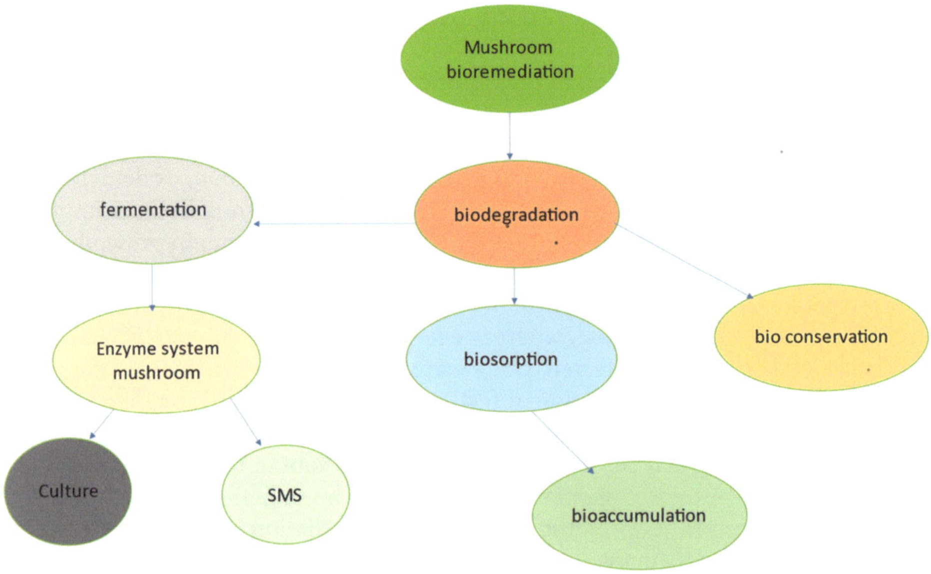

Figure 2.
Steps involved in Mycoremediation [18].

- Fungi being decomposers, decompose dead organic matters.
- Fungi being saprotrophs, feed on dead organic matters.
- Fungal hyphae produce and secrete special acids and enzymes that decomposed lignin (White-rot fungi) and cellulose (brown-rot fungi).
- Fungal mycelium by microfiltration removes toxic substances.
- Fungi are useful in the degradation of oils, petroleum compounds, hydrocarbons, aromatic compounds, and pollutants in soil and water.
- Mushrooms *Agaricus, Amanita, Cortinarius, Boletus, Leccinum, Suillus,* and *Phellinus* are used for mobilization/complexation of different heavy metals in soil [19].

4.2 Phytoremediation

It is a process involving plants for environmental cleanup. There is different process involved as follow; (summarized in **Figure 3**)

- Phytovolatilization: Plants absorb contaminants from the soil and release them into the gaseous atmosphere in an unstable form through the process of transpiration.
- Rhizodegradation: It is the symbiotic relationship between plants and microbes. It is the breakdown of the contaminants due to the presence of protein and enzymes by plants or soil organisms in the rhizosphere.

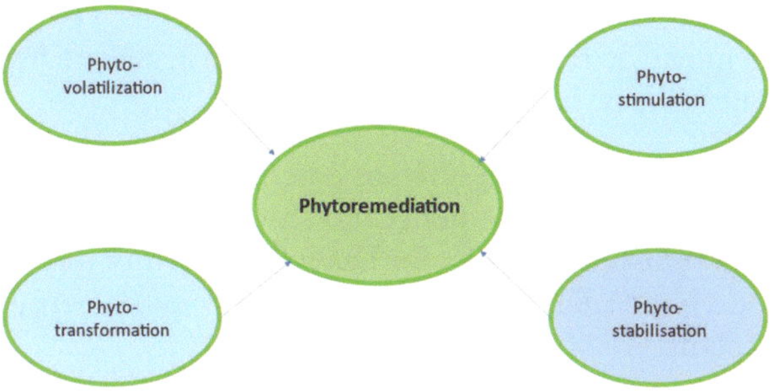

Figure 3.
Components of phytoremediation [20].

- Phytoextraction: Plants take up the contaminants from water and pass them from the roots to the plant's upperparts.

- Phytostabilization: Certain plant species are used to bring contaminants from water and soil.

4.3 Phytoextraction

This technique involves the usage of different algae to extract pollutants from soils, sediments, or water into harvestable plant biomass (hyperaccumulators i.e. those organisms that take larger-than normal amounts of contaminants from the soil). Phytoextraction is more effective for extracting heavy metals than for organic contaminants. The plants translocate contaminants through their root systems to stems and leaves. Different plants absorb different elements and accumulate them into different organs of the plant, as shown in **Figure 4**. For example,

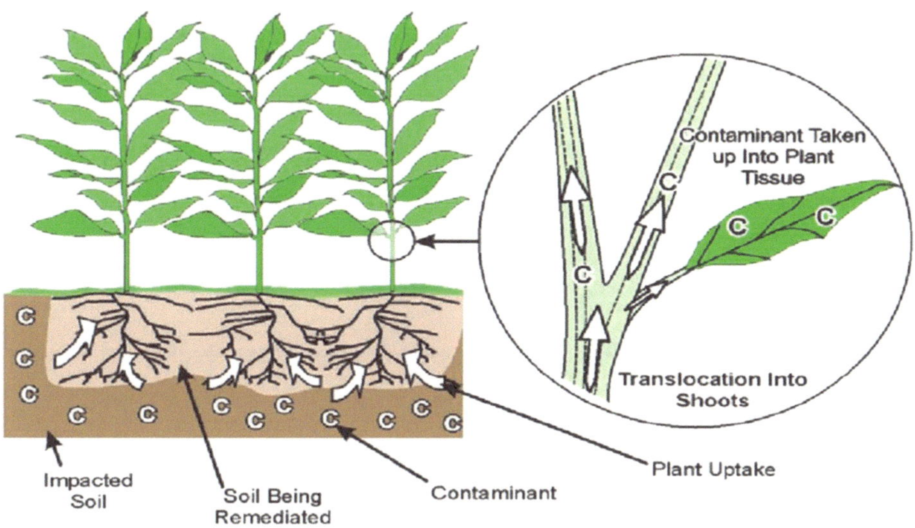

Figure 4.
Phytoextraction for environmental remediation [21].

- Sunflower (*Helianthus annuus*), Chinese Brake fern (*Pteris vittata*) are effectively used for the removal of **Arsenic**. Chinese Brake fern, act as a hyperaccumulator, and accumulates arsenic in its leaves.

- Willow, a common plant, has significant potential as a phytoextractor of **cadmium (Cd), zinc (Zn),** and **copper (Cu).** This plant has some unique characteristics like a high transport capacity for heavy metals from root to shoot and large biomass production. Willow can also be used to produce energy in a biomass-fueled power plant.

- Alpine pennycress (*Thlaspi caerulescens*), a hyperaccumulator, is effective for the removal of metals **Cadmium and Zinc**, although its growth appears to be inhibited by copper.

- Indian Mustard (*Brassica juncea*), Hemp Dogbane (*Apocynum cannabinum*), Ragweed (*Ambrosia artemisiifolia*), or Poplar trees, are useful for the removal of Lead, which sequester lead in their biomass.

- Barley and/or sugar beets, Salt-tolerant (moderately halophytic) varieties are commonly used for the extraction of sodium chloride (common salt) to reclaim saline fields that were previously flooded by high groundwater.

- Selenium, mercury, and other organic pollutants including polychlorinated biphenyls (PCBs) have been removed from soils by transgenic plants containing genes for bacterial enzymes.

4.4 Phycoremediation

It is an excellent form of remediation in aquatic ecosystems. Microalgae, "wonder organisms," are capable of accomplishing bioremediation efficiently by two mechanisms, namely, bioassimilation and biosorption. They have the capability to grow in polluted water as "algal blooms" and assimilate various pollutants. Industrialization has led to increased emission of pollutants into ecosystems. Metal pollutants can easily enter the food chain if heavy metal-contaminated soils are used for the production of food crops.

- Algal blooms have the capability to grow in polluted water and assimilate various pollutants.

- The algal biomass, after harvesting and lipid/protein extraction used as an efficient biosorbent.

- Algal blooms are excellent to remove pesticides from water bodies.

4.5 Bioremediation by microorganisms

Microorganisms are the beneficial source for removing pollutants from soil and water. Microorganisms remove pollutants from water by passive as well as active approaches as shown in **Figure 5**.

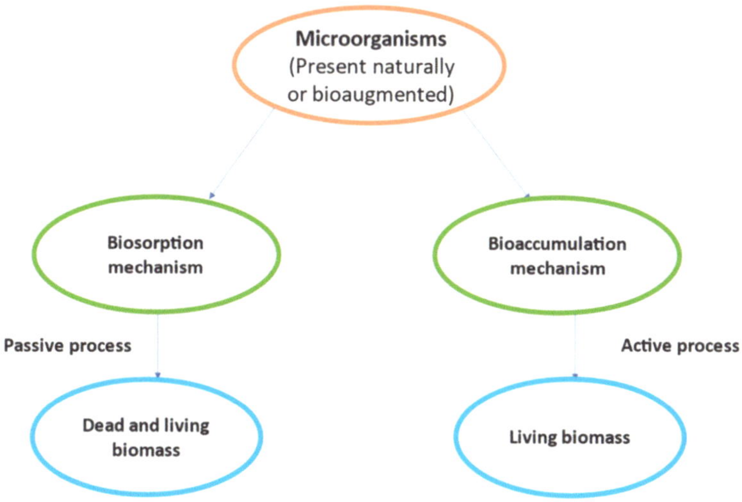

Figure 5.
Steps involved in environmental remediation using microorganisms [22].

- Genetic engineers are working to produce genetically modified microorganisms for bioremediations e.g. *Deinococcus radiodurans*. It is helpful in the absorption of mercury and aromatic hydrocarbons like toluene. Oil spills in water make water unfit and cause the death of life. A large number of marine lives are lost due to these oil spills. Hence causing the disturbance in food chains and ecosystems. These microorganisms are good tools to remediate these oil spills to conserve the environment.

- Protozoa, mites, isopods, and collembolan are used in bioremediation in soil, air, and water.

- Bioaugmentation is the addition of microorganisms to the soil, where biostimulation is the modification, addition, reduction, or genetic engineering of the microbial colonies to degrade pollutants

4.6 Bioremediation by nematodes

Nematode parasites are a sensitive indicator of heavy metals in the aquatic ecosystem showing sharing of more burden of environmental pollution of the sea and also act as bioremediator of heavy metals in fish. In rhizosphere, they are involved in cleaning, nutrient mobilization, nitrification, enzyme activation, etc.

For heavy metal removals, nematodes are being used.

- Bioaccumulation of heavy toxic metals in muscles and guts of fish can be done in the *Echinocephalus* sp. and *Ascaris* sp. which are reported as natural bioremediator of heavy metals in *Liza vaigiensis* [23].

- Nematodes (*Caenorhabditis elegans, Plectus acuminatus, Heterocephalobus pauciannulatus*) are indicators of pollution. They are excellent bioremediators of heavy metals in aquatic habitats.

5. Advantages and disadvantages of bioremediation

5.1 Advantages of bioremediation

The use of naturally available sources for bioremediation increases the efficacy of this process. It imparts socioeconomic as well as environmental benefits to ecosystems.

a. It is a natural waste treatment process. The treatment products are commonly harmless including cell biomass, water, and carbon dioxide.

b. It needs a very less laborious and can commonly carry out on-site, regularly without disturbing normal microbial activities. This also eradicates the transport amount of waste off-site and the possible threats to human health and the environment.

c. It is a cost-effective process in comparison to other conventional methods that are used for clean-up of toxic hazardous waste regularly for the treatment of oil-contaminated sites. It also supports in complete degradation of the pollutants; many of the toxic hazardous compounds can be transformed into less harmful products and disposal of contaminated material.

d. It is chemically benign. Enzymes of microorganisms decontaminate the environment without the addition of toxicants in the environment.

e. This way of remediating the environment is an ecofriendly and economically sustainable approach.

5.2 Disadvantages of bioremediation

Various limitations or disadvantages are associated with bioremediation despite the research and development going on in this field.

a. The process of bioremediation is only applicable to those materials that are biodegradable and cannot be applied as the generalized treatment method for all types of wastes

b. Research is going on to find out more about the persistent and toxic nature of the products of bioremediation

c. As it is a biological process involving the microbial communities so more site-specific environmental specifications are needed for the microorganism to work. In order to maintain that environment, the cost-effectiveness remains there no more.

d. Often it becomes difficult to conduct the pilot-scale bioremediation study in a field study.

e. Genetic engineering of the microbes is needed in order to enhance the efficacy of the bioremediation process [24].

DOI: http://dx.doi.org/10.5772/intechopen.104959

6. Conclusion

Bioremediation is nature's self-healing process by utilizing the hidden workforce capable of decontaminating the environment. The decontaminating ability and efficiency of the biological agents like algae, bacteria, fungi, etc. depends on various factors like oxygen, nutrients, moisture, pH, and temperature. In different regions of the globe, the practice of bioremediation can be made successful by ensuring different factors like cost, the concentration of contaminant, and composition of the degrading site. These factors ultimately ensure the applicability of the ex or in situ bioremediation technique to be implemented. As it has been discussed that ex situ techniques are expensive due to excavation and transportation, although they can treat the large number of contaminants as compared to in situ. The use of various bioremediating agents like bacteria, fungi, algae, and nematodes and further involvement of modern technology including nanoscience is helping to develop new ways to genetically study and engineer the microorganism for need-based functions.

Author details

Amara Dar[1*] and Arooj Naseer[2]

1 Center for Analytical Chemistry, School of Chemistry, University of the Punjab, Lahore, Pakistan

2 Institute of Botany, University of the Punjab, Lahore, Pakistan

*Address all correspondence to: amara.hons@pu.edu.pk

References

[1] Available from: https://microbewiki.kenyon.edu/index.php/File:Bioremediation_images.jpeg

[2] Tiedje JM. Bioremediation from an ecological perspective. In: Situ Bioremediation: When does it work, National Research Council (U.S.), Water Science and Technology Board. Washington, D.C.: National Academy Press; 1993. pp. 110-120

[3] Frutos FJG, Pérez R, Escolano O, Rubio A, Gimeno A, Fernandez MD, et al. Remediation trials for hydrocarbon-contaminated sludge from a soil washing process: Evaluation of bioremediation technologies. Journal of Hazardous Materials. 2012;**199**:262-271. DOI: 10.1016/j.jhazmat.2011.11.017

[4] Smith E, Thavamani P, Ramadass K, Naidu R, Srivastava P, Megharaj M. Remediation trials for hydrocarbon-contaminated soils in arid environments: Evaluation of bioslurry and biopiling techniques. International Biodeterioration & Biodegradation. 2015;**101**:56-65. DOI: 10.1016/j.ibiod.2015.03.029

[5] Frutos FJG, Escolano O, García S, Mar Babín M, Fernández MD. Bioventing remediation and ecotoxicity evaluation of phenanthrene-contaminated soil. Journal of Hazardous Materials. 2010;**183**:806-813. DOI: 10.1016/j.jhazmat.2010.07.098

[6] Sui H, Li X. Modeling for volatilization and bioremediation of toluene-contaminated soil by bioventing. Chinese Journal of Chemical Engineering. 2011;**19**:340-348. DOI: 10.1016/S1004-9541(11)60174-2

[7] Kim S, Krajmalnik-Brown R, Kim J-O, Chung J. Remediation of petroleum hydrocarbon-contaminated sites by DNA diagnosis-based bioslurping technology. Science of the Total Environment. 2014;**497**:250-259. DOI: 10.1016/j.scitotenv.2014.08.002

[8] Firmino PIM, Farias RS, Barros AN, Buarque PMC, Rodríguez E, Lopes AC, et al. Understanding the anaerobic BTEX removal in continuous-flow bioreactors for *ex situ* bioremediation purposes. Chemical Engineering Journal. 2015;**281**:272-280. DOI: 10.1016/j.cej.2015.06.106

[9] Pavel LV, Gavrilescu M. Overview of *ex situ* decontamination techniques for soil cleanup. Environmental Engineering and Management Journal. 2008;7:815-834

[10] Azubuike CC, Chikere CB, Okpokwasili GC. Bioremediation techniques-classification based on site of application: Principles, advantages, limitations and prospects. World Journal of Microbiology and Biotechnology. 2016;*32*(11):180. DOI: 10.1007/s11274-016-2137-x

[11] Available from: http://www.onlinebiologynotes.com/in-siyu-and-ex-situ-bioremediation/ [Accessed April 11, 2022]

[12] Folch A, Vilaplana M, Amado L, Vicent R, Caminal G. Fungal permeable reactive barrier to remediate groundwater in an artificial aquifer. Journal of Hazardous Materials. 2013;**262**:554-560. DOI: 10.1016/j.jhazmat.2013.09.004

[13] Roy M, Giri AK, Dutta S, Mukherjee P. Integrated phytobial remediation for sustainable management of arsenic in soil and water. Environment

International. 2015;**75**:180-198. DOI: 10.1016/j.envint.2014.11.010

[14] Frascari D, Zanaroli G, Danko AS. In situ aerobic cometabolism of chlorinated solvents: A review. Journal of Hazardous Materials. 2015;**283**:382-399. DOI: 10.1016/j.jhazmat.2014.09.041

[15] Manisha M, Sandeep KS, Ajay K. Environmental factors affecting the bioremediation potential of microbes (chapter 5). In: Woodhead Publishing Series in Food Sciences, Technology and Nutrition. 2021. pp. 47-58. DOI: 10.1016/b978-0-12-821199-1.00005-5. (ISBN 9780128211991)

[16] Bharath Y, Singh SN, Keerthiga G, Prabhakar R. Mycoremediation of contaminated soil in MSW sites. In: Ghosh SK, editor. Waste Management and Resource Efficiency. Singapore: Springer Nature; 2019. pp. 321-329. DOI: 10.1007/978-981-10-7290-1_28

[17] Treu R, Falandysz J. Mycoremediation of hydrocarbons with basidiomycetes—A review. Journal of Environmental Science and Health, Part B. 2017;**52**(3):148-155. DOI: 10.1080/03601234.2017.1261536

[18] Ali A, GuoO D, Mahar A, Wang P, Shen F, Li R, et al. Mycoremediation of potentially toxic trace elements—A biological tool for soil cleanup: A review. Pedosphere. 2017;**27**(2):205-222. DOI: 10.1016/S1002-0160(17)60311-4

[19] Mushroom mycoremediation: kinetics and mechanism - Scientific Figure on ResearchGate. Available from: https://www.researchgate.net/figure/Mechanism-of-mycoremediation_fig1_333661657 [Accessed February 22, 2022]

[20] Phytoremediation: An Ultimate Hope for the Planet - Scientific Figure on ResearchGate. Available from: https://www.researchgate.net/figure/Mechanism-of-Phytoremediation_fig1_301620916 [Accessed February 22, 2022]

[21] The Feasibility of Poplars for Phytoremediation of TCE Contaminated Groundwater: A Cost-Effective and Natural Alternative Means Of Groundwater Treatment - Scientific Figure on ResearchGate. Available from: https://www.researchgate.net/figure/2-Phytoextraction-Process-This-figure-depicts-the-phytoextraction-process-in-plants_fig4_265191596 [Accessed February 22, 2022]

[22] Kiran B, Pathak K, Kumar R, Deshmukh D. Phycoremediation: An eco-friendly approach to solve water pollution problems. In: Kalia V, Kumar P, editors. Microbial Applications. Vol. 1. Cham: Springer; 2017. DOI: 10.1007/978-3-319-52666-9_1

[23] Azmat R, Fayyaz S, Kazi N, Mahmood SJ, Uddin F. Natural bioremediation of heavy metals through nematode parasite of fish. Biotechnology. 2008;**7**(1):139-143

[24] Available from: Advantages And Disadvantages of Bioremediation 4 Feb.2018 (agriculturistmusa.com) [Accessed April 14, 2022]

Chapter 4

Biological Treatments for Petroleum Hydrocarbon Pollutions: The Eco-Friendly Technologies

Innocent Chukwunonso Ossai, Fauziah Shahul Hamid and Auwalu Hassan

Abstract

Anthropogenic activities introduce petroleum hydrocarbons into the environments, and the remediation of the polluted environments using conventional physicochemical, thermal, and electromagnetic technologies is a challenging task, laborious work, and expensive. The ecotoxicological effects and human health hazards posed by petroleum hydrocarbon pollutions gave rise to the call for "green technologies" to remove petroleum hydrocarbon contaminants from polluted environments. It is imperative to transition from the conventional physicochemical treatments methods that are expensive to more eco-friendly biological treatment technologies that reduce energy consumption, chemicals usage, cost of implementation and enables more sustainable risk-based approaches towards environmental reclamation. The chapter summarises and gives an overview of the various biological treatment technologies adapted to the remediation of hazardous petroleum hydrocarbon polluted sites. Biological treatment technologies include; bioremediation, biostimulation, bioaugmentation, bioattenuation, bioventing, biosparging, bioslurry, biopiling, biotransformation, landfarming, composting, windrow, vermiremediation, phytoremediation, mycoremediation, phycoremediation, electrobioremediation, nanoremediation, and trichoremediation. They are green technology approaches widely adopted, scientifically defensible, sustainable, non-invasive, ecofriendly, and cost-efficient in the remediation of petroleum hydrocarbons polluted environments compared to the physicochemical, thermal, and electromagnetic treatments technologies, which are rather destructive and expensive. The chapter provides detailed illustrations representing the various biological treatment technologies for a comprehensive understanding and successful implementation with their subsequent benefits and constraints.

Keywords: bioremediation, phytoremediation, phycoremediation, mycoremediation, vermiremediation, trichoremediation

1. Introduction

The intensive development of human civilisation, urbanisation, population growth, economic development, and impulsive industrialisation have expanded petroleum hydrocarbon production, distribution, and utilisation. This phenomenon caused a gradual depletion of natural petroleum reserves and increasing demand for petroleum products [1]. The petroleum industry is one of the world's largest and most important global industries with a primary function in oil and gas production [2]. The global economy has become entangled with infrastructure that depends on petroleum hydrocarbon products such as petrol, diesel, kerosene, jet fuel, fuel oil and motor oils [3]. These products have become the main source of primary energy globally. Their exploration has transformed the world by providing fuel and raw materials for various industries for various applications and serving as feedstock for several consumer goods, thus playing an increasing and relevant role in our daily lives [4]. Apart from the benefit of being an important energy source, the products have caused the environment to become constantly bombarded with hazardous pollutants [5]. The causes of the pollutants entering the environment are diverse (**Figure 1**) as the amount of individual petroleum hydrocarbon components are significantly substantial. Pollution caused by petroleum hydrocarbon products poses direct and indirect ecotoxicological effects and human health risks [6–8].

The environmental fate and toxicokinetics of petroleum hydrocarbons are critical aspects of risk assessment because they determine human or environmental receptor exposure to pollution [9, 10]. When discharged or released in the environment, the components of petroleum hydrocarbons undergo weathering processes [11], involving various processes such as adsorption, volatilisation, dissolution, biotransformation, photolysis, oxidation, hydrolysis through interaction with microorganisms and metabolic pathways [12, 13]. The level at which various components of petroleum hydrocarbon deteriorate under weathering processes depends mainly on the nature of the petroleum hydrocarbon compounds, composition, physical and chemical characteristics [14]. A wide variety of natural processes involved in the fate and behaviour of petroleum hydrocarbons in the soil are illustrated in **Figure 2**. The weathering process includes adsorption to soil particles and organic materials, volatilisation to the atmosphere [15], and dissolution in water [16]. Environmental conditions, such as temperature, humidity

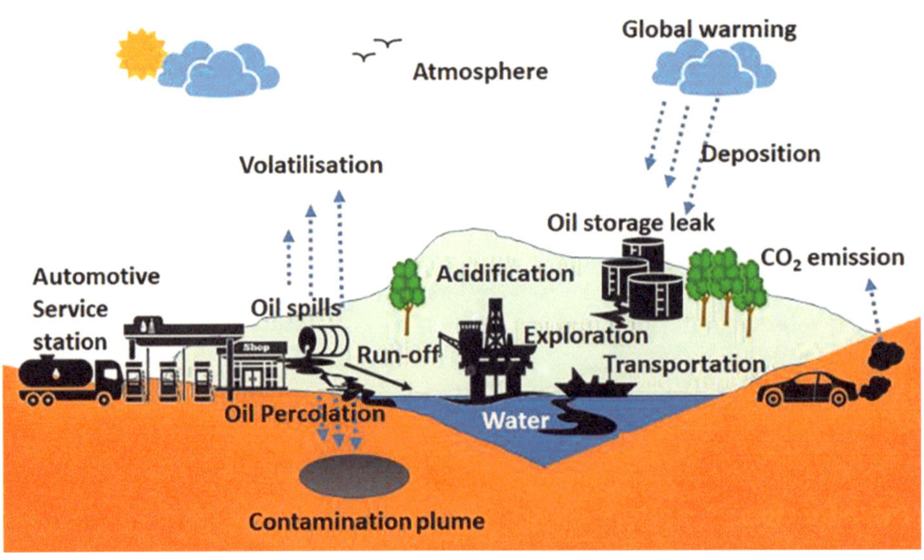

Figure 1.
Sources of petroleum hydrocarbon pollution.

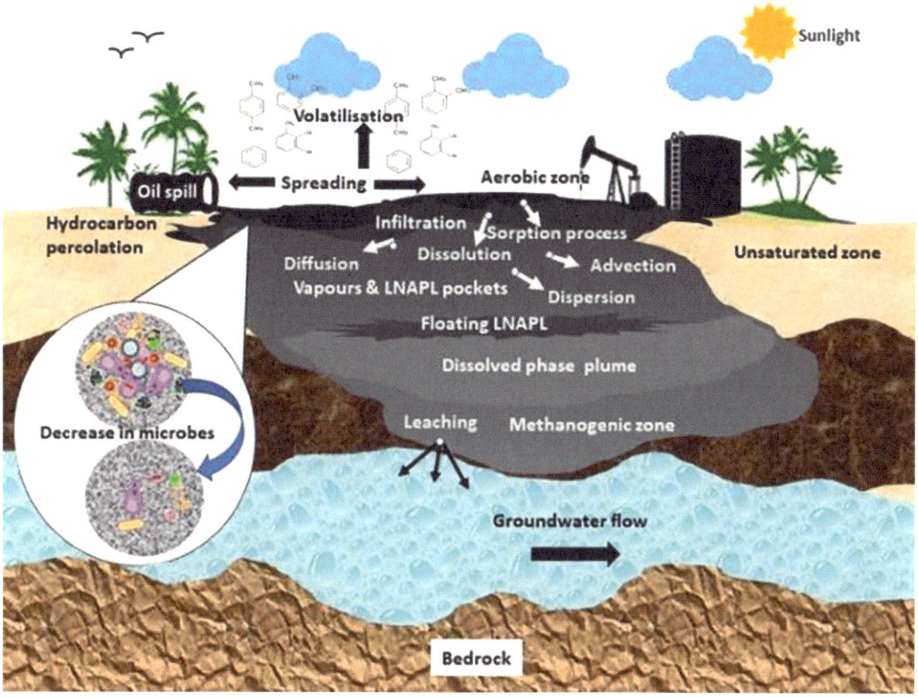

Figure 2.
Environmental fate of petroleum hydrocarbon on soil [11].

and precipitation, affect the weathering process [11]. The aliphatic hydrocarbons are more readily biodegraded than aromatic hydrocarbons [17], and the aliphatic hydrocarbons are more volatile because of their molecular nature [18]. If volatilisation is the primary weathering process, the loss of lower molecular weight aliphatic hydrocarbons is the most dominant change in the petroleum hydrocarbon, which may be the principal air pollutants causing air pollution at contaminated sites [19]. Volatilisation changes the residual non-aqueous liquid (NAL), affecting its transportation over time [20]. The petroleum hydrocarbon vapours are transported to the gaseous phase through diffusion or advection, and the process depends on the soil pore characteristics [21]. The gas-phase mass transfer in a polluted soil consists of volatilisation from the non-aqueous phase liquid (NAPL) and partitioning in gaseous/aqueous interphase [14].

However, considering the environmental impacts of petroleum hydrocarbons which affect the surface soil, subsoil, sediments, surface water and groundwater coupled with the human health risk. It has become imperative to transition from conventional treatment technologies such as physicochemical treatments, thermal/heat treatments, electric and electromagnetic treatments, acoustic and ultrasonic treatments that are challenging, laborious, extensive and expensive to more feasible biological treatment technologies that are sustainable, eco-friendly and economical.

2. Biological treatment technologies

Biological treatment technologies that have shown remarkable success for *in situ* and *ex situ* remediation of petroleum hydrocarbons are illustrated in **Figure 3**.

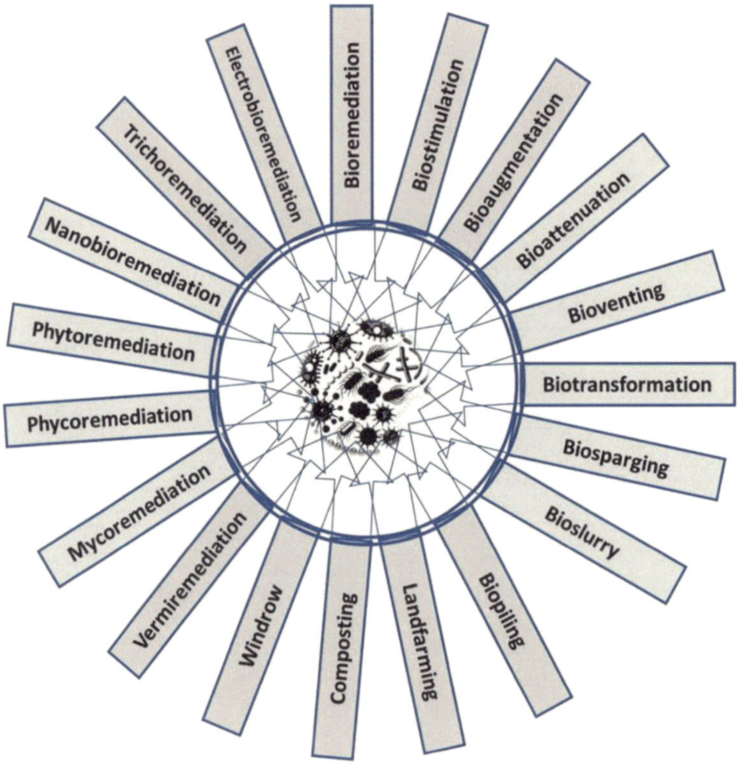

Figure 3.
The biological treatment technologies for petroleum hydrocarbon remediation.

The feasibility of the biological treatment technology depends mainly on the limiting factors and the location of the contaminants. Treatability also depends on the soil, sediments, surface water, and groundwater properties, whether it is localised or removed, excavated, and transported for treatment at an off-site treatment facility. If treatment is on-site, the term *in situ* suffices, and if treatment is off-site, *ex situ* suffices [22]. The biological treatment technologies can remediate or degrade petroleum hydrocarbons and various organic contaminants to simpler and non-toxic substances without any long-term adverse effect on the impacted environments [23]. The general advantage of biological treatment technologies is that treatments do not disrupt the environment. The general constraint is that treatments usually require a long treatment period ranging from months to several years for a satisfactory and effective removal of contaminants. High concentrations of contaminants may result in low microbial activity with low or insufficient removal efficiency [24].

2.1 Bioremediation

Bioremediation is an eco-friendly, sustainable, and cost-effective means of restoring and cleaning soil contaminants such as petroleum hydrocarbons in polluted environments. The technique comprises the natural degradation of petroleum hydrocarbon contaminants by petroleum hydrocarbon-degrading microorganisms such as bacteria, fungi, yeasts, and algae. Bioremediation removes and neutralises hazardous petroleum hydrocarbon contaminants to non-toxic or simpler compounds such as carbon (IV) oxide and water through oxidation process under aerobic conditions by the microorganisms with the nutrient provision and optimisation of the constraining factors for efficient metabolic activities [25, 26]. The petroleum hydrocarbon-degrading microorganisms in the soil participate in defining the metabolic pathways and mechanisms of the microbial degradation of petroleum hydrocarbons [27]. Bioremediation of alkanes typically occurs via a sequential oxidation process by a few microbial enzymes (i.e., alkane monooxygenases or cytochrome P450 oxidases, alcohol dehydrogenases, and aldehyde dehydrogenases) and connects to the cytosolic fatty acid metabolism (**Figure 4**).

Some genes affiliated with the outset of petroleum hydrocarbon metabolism have been identified, as *alk*B (encoding alkane monooxygenase) and *ndo* (encoding naphthalene dioxygenase). These genes are activated under aerobic conditions to degrade alkanes and polycyclic aromatic hydrocarbons (PAHs), respectively [28]. Before implementing the bioremediation, it is essential to consider all the limiting factors such as energy sources, pH, temperature, nutrients and inhibitory substances, which may affect the success of the bioremediation process [29]. In bioremediation, the aliphatic petroleum hydrocarbons are more amendable or degradable by the microorganisms than the long-chain and the branched or cyclic chain petroleum hydrocarbons [19]. The petroleum hydrocarbon-degrading microorganisms utilise carbon compounds as energy sources, growth, and reproduction [30]. Bioremediation using selected microorganisms or genetically modified microorganisms is increasing the interest of many researchers.

Some of the most commonly isolated petroleum hydrocarbon-degrading bacteria belong to the genus *Acinetobacter*, *Alcaligenes*, *Paenibacillus*, and *Pseudomonas* [31] and are recognised to efficiently degrade hazardous petroleum hydrocarbon contaminants into simpler compounds [32, 33]. In addition, fungi species such as *Penicillium*, *Fusarium*, and *Rhizopus* have been isolated and utilised in the bioremediation of petroleum hydrocarbon contaminated soil and sediments [34, 35]. However, bioremediation of petroleum hydrocarbon has been in use since 1940 but gained popularity after the Exxon Valdez spill in 1980 [36]. Bioremediation has been successfully

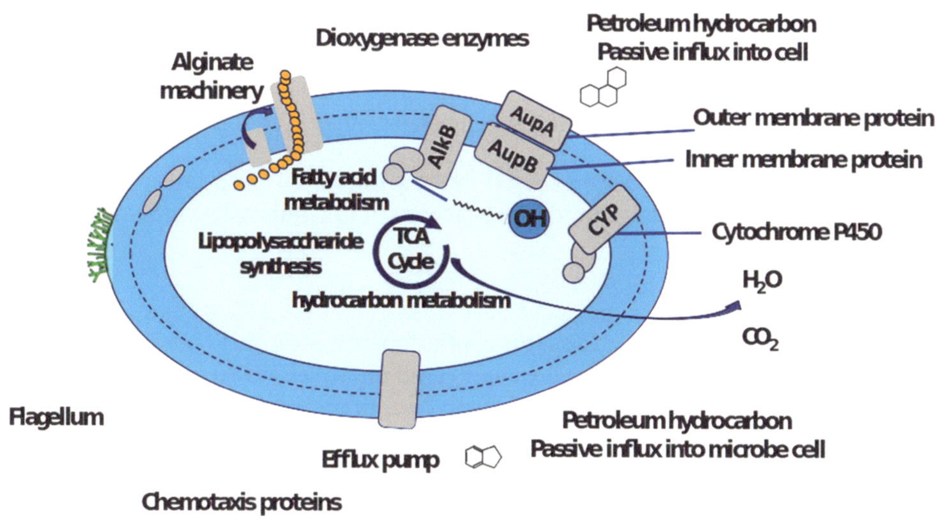

Figure 4.
Microbial bioremediation of petroleum hydrocarbon [27].

applied worldwide in environmental oil pollution mitigation, such as in the oil spills in Prince William Sound, Alaska, in 1989 [37] and the Gulf of Mexico in 2010 [38], and it is a promising strategy for environmental cleanup in contaminated mangrove sediments [28, 39].

The advantages of bioremediation include; minimal disruption of the ecosystem, permanent elimination of contaminants, cheap operation costs, and can be coupled with other treatment technologies. The disadvantages include extensive monitoring, production of unknown by-products, long duration to complete bioremediation, and bioremediation limited to biodegradable compounds [40].

2.2 Biostimulation

Biostimulation involves adding stimulatory materials, organic wastes (**Figure 5**), bulking agents, nutrients amendments, bio-surfactants, biopolymers, and slow-release fertilisers to enhance and support microbial growth and enzymatic activities of the indigenous microorganisms in the contaminated soil for remediation activities [23, 41, 42].

Biostimulation occurs by optimising various rate-limiting parameters such as pH, temperature, aeration, macromineral nutrients, and electron acceptors such as carbon, oxygen, nitrogen, phosphorus, and potassium, which accelerate the metabolic activities of the indigenous microorganisms [43]. Biostimulation can be performed *in situ* and *ex situ* but depends on the existence of the indigenous microorganisms with the capacity to degrade the hazardous contaminants [44, 45]. The microbial community composition becomes evener and richer during biostimulation [46], and the requirements include the presence of correct microorganisms, ability to stimulate target microorganisms, ability to deliver nutrients, C:N:P-30:5:1 for balance growth [45]. A study conducted by Singh et al. [47] investigated biostimulation of petroleum hydrocarbon contaminated soil using bacterial consortia and nutrient mixture to achieve a TPH removal efficiency of 99.9% after 18 months.

The benefits of biostimulation include; the use of native microorganisms adapted to the environment, being eco-friendly and cost-effective, preventing ecosystem disturbance, and can be coupled with other treatment technologies. The disadvantages include; it depends on environmental factors that control the potentiality, requiring extensive monitoring and scientific observations, contaminants may be non-biodegradable after adsorption to soil particles, and it takes a long duration to complete degradation [48, 49].

Various organic wastes have been used for biostimulation to optimise the degradation and removal of total petroleum hydrocarbons in the polluted soil [50–52].

2.3 Bioaugmentation

Bioaugmentation involves adding exogenous microbial cultures, autochthonous microbial communities, or genetically engineered microbes with a specific catabolic activity that have adapted and proven to degrade contaminants to enhance degradation or increase the rate of degradation of contaminants [17, 53–55]. Alexander [56] described bioaugmentation as inoculating contaminated soil or sediments with specific strains or consortia of microorganisms to degrade pollutants in the soil. Soil microbial community composition changes while microbial diversity decreases by bioaugmentation treatment [46].

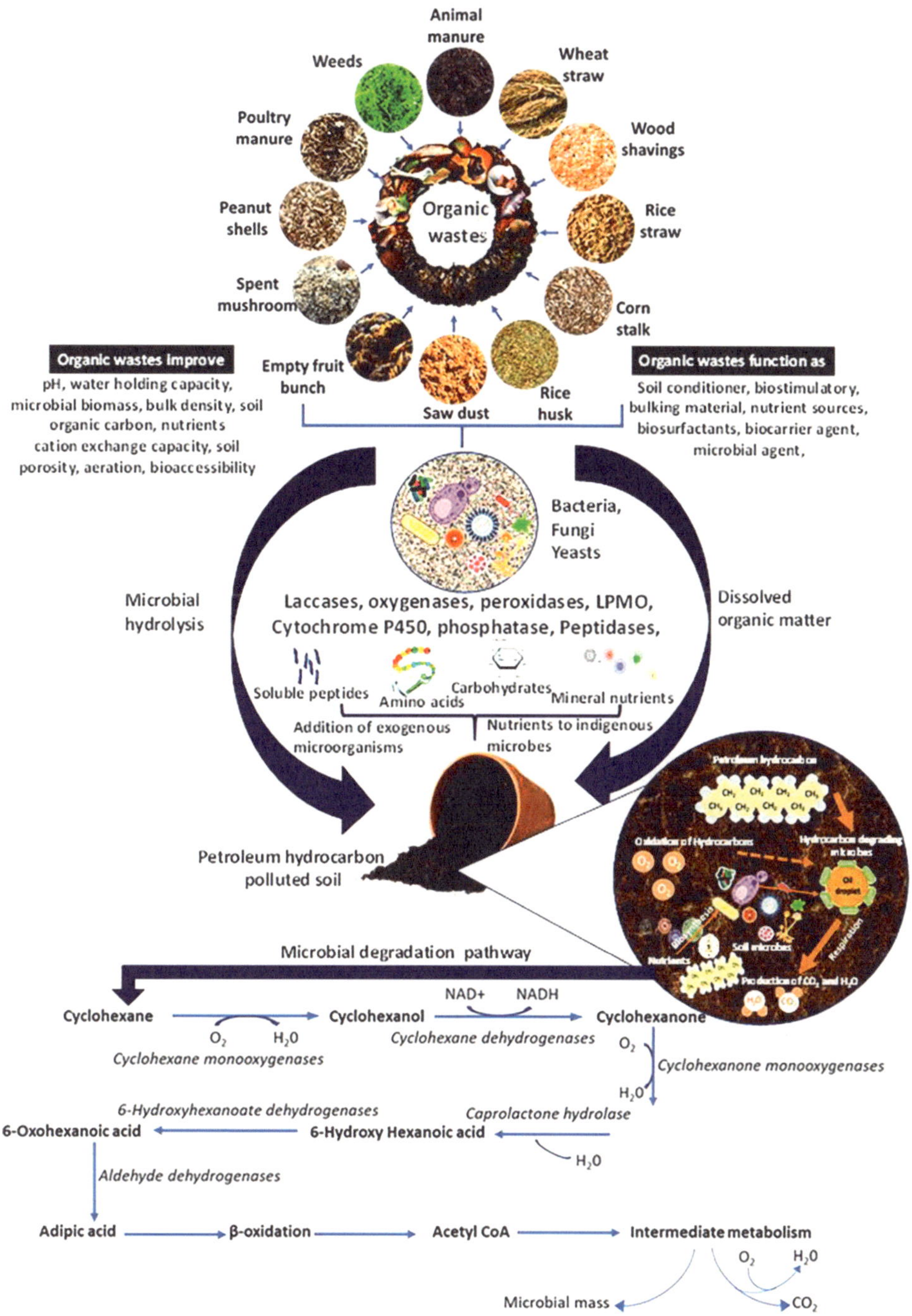

Figure 5.
Organic wastes used in biostimulation of petroleum hydrocarbons.

Genetically engineered microorganisms have shown potential in bioaugmentation, exhibiting enhanced degrading capabilities for broad coverage of chemical and physical pollutants [57]. In the oil-polluted site of ONGC field in Gujarat, India, Varjani et al. [58] demonstrated *in situ* bioaugmentation using hydrocarbon utilising bacteria consortium comprising six bacterial isolates for degradation of petroleum hydrocarbon contaminants and achieved removal efficiency of 83.7% in 75 days. Corvino et al. [59] also demonstrated bioaugmentation by using autochthonous fungi from petroleum hydrocarbon contaminated soil to degrade clay soil contaminated with petroleum hydrocarbons and achieve a removal efficiency of 79.7% after 60 days period.

The benefits of bioaugmentation include; less labour demand, the microbes do the work once introduced, microbial strains, mixed cultured or indigenous microbes can be used, eco-friendly and cost-effective, and can be carried out *in situ* without soil excavation. It can be combined with other treatment technologies. The disadvantages of bioaugmentation include; microbes require an appropriate environmental condition to thrive, the microbes may not metabolise all the contaminants completely, indigenous microbes may outcompete the introduced microbes, long duration to complete the remediation and may require genetically engineered microbes for degradation of contaminants [60].

2.4 Bioattenuation

Bioattenuation or natural attenuation is the use of naturally occurring processes, including a variety of physical and biochemical processes without human intervention, to remove, transform, neutralise and reduce the mass, volume, concentration, and toxicity of hazardous contaminants such as petroleum hydrocarbons in the environment by the activities of the indigenous microorganisms [28]. The process occurs through advection, dispersion, sorption, dissolution, volatilisation, chemical transformation, abiotic and biological transformation, stabilisation, and biodegradation [42]. Bioattenuation is applicable for contaminated environments with low contaminant concentrations and used in places where other remediation methods cannot be adopted [61].

The benefits of bioattenuation include; it can be adopted in all areas, causes minimal disruption of the site and the environment, low cleanup cost and can be used in conjunction with or as a follow up to other remediation methods. The disadvantages include; it is not all contaminants that are susceptible to rapid and complete degradation, it requires extensive site monitoring over a long period, it is limited to biodegradable contaminants, it depends on environmental factors that control potentiality for its success, and bioattenuation alone is inadequate and protracted in many cases [62].

2.5 Bioventing

Bioventing is an *in situ* bioremediation technology that utilises the indigenous microorganisms to biodegrade hazardous organic pollutants adsorbed to the soil. The technique involves injecting air (oxygen) into the contaminated soil to increase the *in situ* degradation and minimise the emission of volatile contaminants to the atmosphere [63, 64]. The injection of air into the soil stimulates and increases aerobic conditions for the growth of indigenous microorganisms and enhances the catabolic activity of the contaminants [65]. The mechanism of the bioventing process is similar to soil vapour extraction. Soil vapour extraction removes volatile pollutants through volatilisation, while bioventing systems promote biodegradation and minimise volatilisation [66]. Bioventing is helpful in the remediation of petroleum hydrocarbon contaminated soil. A bioventing layout using extraction vent wells is illustrated in **Figure 6**.

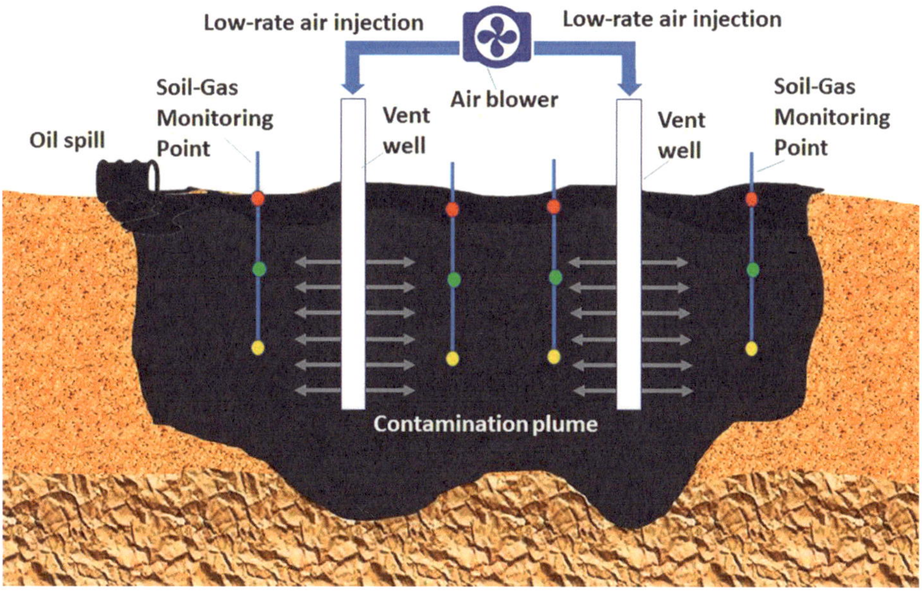

Figure 6.
Bioventing system for remediation of polluted soil [67].

In a bioventing system study conducted by Agarry and Latinwo [42], the bioventing process was demonstrated on diesel oil-contaminated soil amended with brewery effluents as an organic nutrient source and achieved a removal efficiency of 91.5% over 28 days period. A similar study by Thomé et al. [68] also assessed the bioventing process on diesel-contaminated soil without any soil amendment and obtained a removal efficiency of 85% after 60 days.

The benefits of bioventing include; it can be deployed for *in situ,* and *ex situ* cleanup of contaminants, causes minimal disruption of the environment, low cleanup cost, and can be used in conjunction with other treatment technologies or as a follow up to other remediation methods. The disadvantages include; it does not promote remediation when the contamination zone is anaerobic, difficult to minimise environment release, low permeability soil pose a challenge due to its limited ability to distribute air through the surface, lab-scale and pilot-scale cannot guarantee treatment standards for specific contaminants of concern. Bioventing alone is inadequate and protracted in many cases [69].

2.6 Biotransformation

Biotransformation is a biotechnological process that involves modifications in the chemical constituents of the hazardous pollutants by the microorganisms or enzyme-mediated systems to form molecules with high polarity [70]. The mechanism transforms organic compounds from one form to another to reduce the contaminants' toxicity and persistence [71, 72]. Naturally, the biotransformation process occurs very slowly and is nonspecific and less productive. But microbial biotransformation or biotechnology generates high amounts of metabolites, more rapid and productive outcomes, with more specificity. Microbial biotransformation helps modify and transform various contaminants and a large variety of compounds, including petroleum hydrocarbons in the soil [69]. Biotransformation of petroleum hydrocarbon contaminated soil occurs through bacteria, fungi, and yeast metabolic activities [38]. However, genetically modified organisms (GMOs) or genetically engineered microorganisms (GEMs) have shown potential in the biotransformation of contaminants in soil [57]. Biotransformation processes occur through oxidation, reduction, denitrification, condensations, isomerisation, hydrolysis, sulphidogenesis, methanogenesis, functional group introduction, and new bonds, as illustrated in **Figure 7** [73].

In a pilot-scale investigation, Al-Bashir et al. [74] demonstrated a biotransformation study of naphthalene at the concentration of 50 mg/L in a slurry system under denitrifying conditions for 50 days. The results indicated that 90% of the total naphthalene was transformed after 50 days at a maximum mineralisation rate of 1.3 mg L^{-1} per day.

The benefits of biotransformation include; it can be deployed for *in situ* and *ex situ* cleanup processes, uses microbial enzymes to metabolise contaminants and causes less disruption of the site and the environment. The disadvantages include; it may constitute cost due biotechnological process to synthesise biocatalysts, biosurfactants and enzymes, the contaminants may inhibit or kill the microbes, efficiency depends on the quality of the biocatalysts produced by microbes, required extensive biomonitoring and assessment, and required modification of microbes to produce target biocatalysts [69].

2.7 Biosparging

Biosparging involves the injection of air (oxygen) and nutrients into the saturated zone under pressure to increase groundwater oxygen concentration to stimulate biological activities of the indigenous microorganisms to degrade contaminants [67, 75].

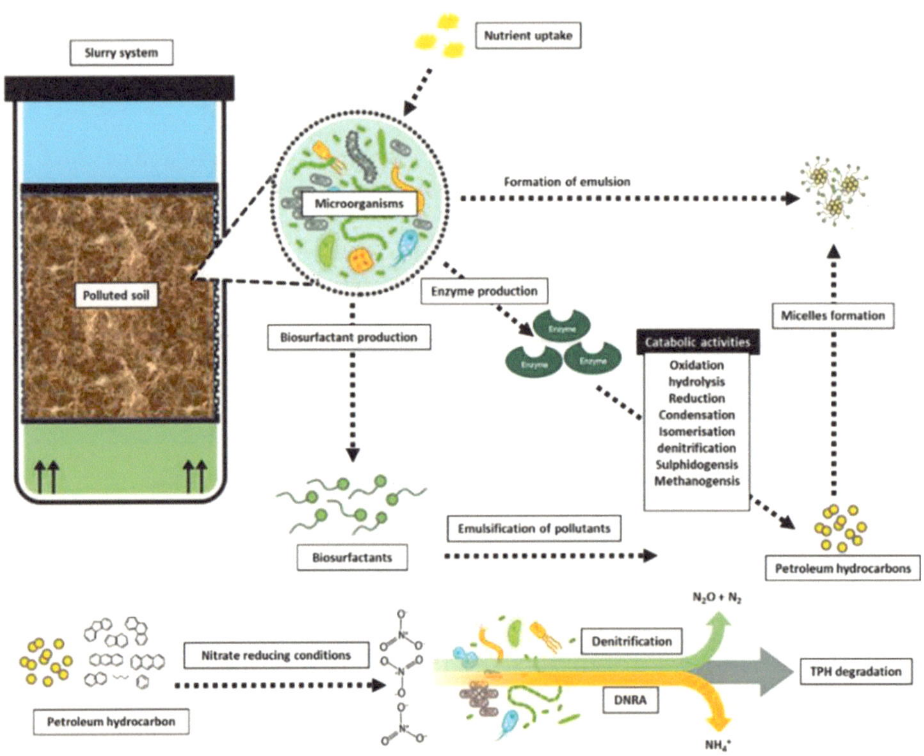

Figure 7.
Biotransformation mechanism under the denitrifying conditions.

Biosparging technology helps to reduce the contaminant concentration adsorbed to the soil, within the capillary fringe above the water table, and contaminants dissolved in the groundwater. The effectiveness of biosparging depends on soil permeability and pollutant degradability [76]. **Figure 8** illustrates the biosparging process in a polluted site.

In a study conducted by Kao et al. [78], a biosparging technique was deployed in a petroleum oil spill site for 10 months, and the result produced 70% removal efficiency for benzene, toluene, ethylbenzene and xylene (BTEX) within the remedial period.

The benefits of biosparging include; the equipment is easy to instal, creates minimal disturbance to site operation, requires no soil removal or excavation, and a low air injection rate minimises the potential need for vapour capture, and treatment is cost-competitive. The limitation of biosparging is in predicting the direction of airflow in the process as it depends on the high airflow rate to achieve pollutant volatilisation and promote degradation [79]. It is site-specific and can cause the migration of contaminants, some interactions among complex chemicals and biophysical processes are not well understood and used only where suitable [66].

2.8 Bioslurry

Bioslurry involves the treatment of contaminated soil in a controlled bioreactor such as sequencing batch, feed-batch, continuous and multistage bioreactors [80, 81]. In a bioslurry treatment system, nutrients are added to enhance microbial activities to degrade hazardous contaminants. The bioslurry reactor is designed with various process controls to monitor, control, and manipulate temperature, mix, and add nutrients to achieve maximum removal efficiency. Amendments such as designer bacteria, surfactants, and enzyme inducers can be used in slurry bioreactors to stimulate and enhance biodegradative activities [82]. Bioslurry reactors may be constructed to provide sequential anaerobic/aerobic treatment conditions, as illustrated in **Figure 9**.

Bioslurry is an *ex situ* technology that can be used for bioremediation of problematic sites (when the less expensive natural attenuation or stimulated *in situ* bioremediation are not feasible [84]. The technology has been applied only to remove substances that are not readily degradable and non-halogenated volatile organic compounds, petroleum hydrocarbons and explosive compounds. Slurry-phase bioreactors

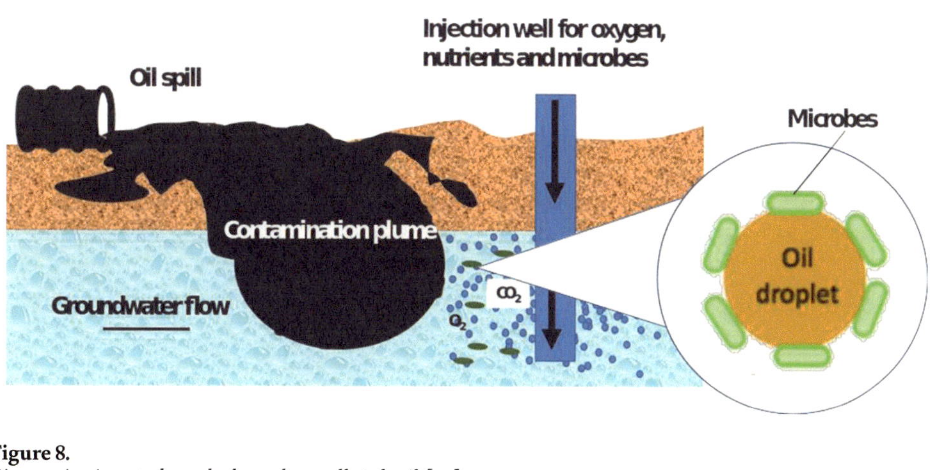

Figure 8.
Biosparging in petroleum hydrocarbon polluted soil [77].

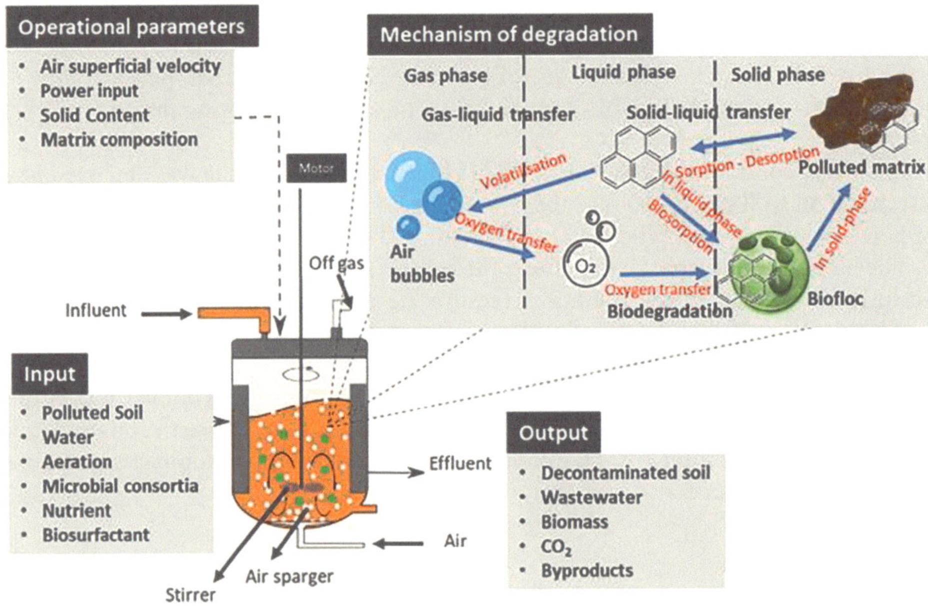

Figure 9.
Bioslurry mechanisms [83].

containing co-metabolites and specially adapted microorganisms are used *ex-situ* to treat halogenated compounds, pesticides, polychlorinated biphenyls (PCBs) [85].

In a study conducted by Tuhuloula et al. [86, 87], bioslurry treatment was demonstrated on petroleum hydrocarbon contaminated soil obtained from the oil drilling site of Pertamina Petrochina in Indonesia using microbial consortia of *Bacillus cereus* and *Pseudomonas putida*. The result obtained showed naphthalene removal efficiency between 79.35–99.73% in a slurry bioreactor after 49 days. A similar pilot-scale study conducted by Zhang et al. [85] evaluated aerobic bioslurry phase reactors in treating soil contaminated with explosive compounds (2,4 and 2,6-dinitrotoluenes) at Army Ammunition Plant in Tennesse and Wisconsin, USA. The result obtained showed a removal efficiency of 99%.

The benefits of bioslurry-phase treatment include increased intimated contact between microorganisms and the contaminants, faster degradation rate more than other biological treatments, provides greater control of environmental and operating conditions, and gas emissions are controlled and harnessed as biogas and requires small site space. The disadvantages include; it is an *ex situ* process and requires soil excavation, dewatering of soil after treatment is required and can be expensive, the treatment cost is high when off-gas is treated due to volatile compounds, and sizing materials is difficult and expensive as non-homogeneous soil and clayey soil create materials handling issues, and further treatment of non-recycled effluent is required [82].

2.9 Landfarming

Landfarming, also known as land treatment or land application, is an above-ground form of bioremediation technology that involves engineered bioremediation systems that employ tilling, ploughing, and spreading the polluted soil in a thin layer on the land surface to enhance and stimulate aerobic microbial activities with the addition of nutrients, mineral and moisture to reduce the pollutant level biologically [86]. It is suitable for treating soil contaminated with low molecular weight petroleum hydrocarbons, volatile organic compounds (VOCs), and other organic compounds [88]. Enhancing biodegradation in landfarming is achieved by adding oxygen, moisture and nutrients [89]. Tilling also introduces oxygen to the soil and helps increase evaporation while adding nutrients or soil amendments such as organic wastes or organic fertilisers provide nutrients to stimulate microbial activities [90]. **Figure 10** illustrates the component in the landfarming system for petroleum hydrocarbon contaminated soil.

The Landfarming method has been proven effective in reducing all the constituents of petroleum hydrocarbons at underground storage tanks. Low molecular hydrocarbons tend to be removed by volatilisation during landfarming aeration, tilling and ploughing and degraded through microbial respiration. The heavy molecular hydrocarbons do not volatilise during landfarming aeration but undergo breakdown by biodegradation activity by the soil microorganism [66].

The study demonstrated by Brown et al. [88] showed landfarming to improve biological treatment of petroleum hydrocarbons in the soil in 110 days with nutrient addition. The results obtained after 6 weeks showed 53% for total petroleum hydrocarbon (TPH) removal from the contaminated soil. Landfarming is a successful treatment option for remediation of petroleum hydrocarbon contaminated soil.

The benefits of landfarming treatment include; low capital input, simple technology design and implementation, a large volume of polluted soil can be treated, *in situ* and *ex situ* application, negligible environmental impact and energy efficiency. The disadvantages include; it is limited to removal of biodegradable pollutants, a large treatment area is required, involves pollutant exposure risks, excavation incurs additional cost, and it provides limited knowledge of the microbial process or the unravelling limitation factors during remediation [91].

2.10 Bio-piling

Bio-piles, also known as bio-cells, bio-heaps, bio-mounds and compost piles, are used to reduce the concentrations of hazardous petroleum hydrocarbon contaminants

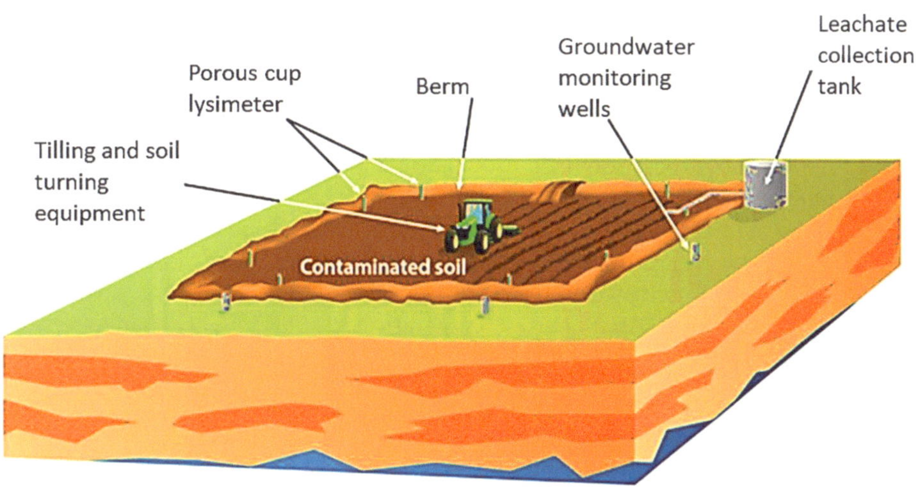

Figure 10.
Landfarming of contaminated soil [86].

in excavated soils through biodegradation. The technology involves a combination of landfarming and composting in an engineered cell aerated with blowers and vacuum pumps, irrigation and nutrient system, and leachate collection system for bioremediation of pollutant components adsorbed to soil and sediments [92]. The technique involves piling an excavated contaminated soil, followed by biostimulation and aeration to enhance microbial activities for degradation [93]. It is suitable for treating a large volume of contaminated soil and sediments in a limited space and effectively remedy pollutions in extreme environments [94, 95].

The essential components of the technique include the addition of air (oxygen), moisture (water), nutrients and bulking agents (organic materials), leachate collection system and treatment bed [96]. Biopiling of contaminated soil can limit the volatilisation of low molecular weight contaminants in petroleum hydrocarbons [97]. Biopile systems are similar to landfarms in that they are both engineered and aboveground systems that use oxygen from the air to stimulate the growth and reproduction of aerobic microorganisms, which degrade the adsorbed petroleum hydrocarbon contaminants in the soil. While landfarms are aerated through tilling or ploughing, biopiles are aerated through air injection or extraction through slotted or perforated piping placed throughout the piles [66]. **Figure 11** illustrates the biopiling process for remediation of petroleum hydrocarbon contaminated soil.

Gomez and Sartaj [98] demonstrated a study by conducting biopiling treatment of petroleum hydrocarbon contaminated soil at a low-temperature field scale using consortia of microorganisms and organic compost for 94 days. The result obtained showed a removal efficiency of 90.7% for total petroleum hydrocarbon (TPH).

The benefits of biopiling include; it is relatively simple to design and implement, effective for pollutants with slow biodegradation rates, it can be designed to be a closed system with vapour emission controls, it requires less land area than landfarms, and cost-effective. The limitations include; contaminants reduction >95% and concentration <0.1 ppm are challenging to achieve, not practical for high pollutant concentrations, volatile compounds tend to evaporate rather than biodegrade during treatment, a large land area is required, vapour generation require treatment before discharge, and requires bottom liners to prevent leaching [66].

2.11 Composting

Composting is a controlled microbial aerobic biochemical degradation of organic waste materials and its conversion into a stabilised organic material that can be useful as soil conditioners for remediation of soil contaminated with organic compounds such as petroleum hydrocarbons [99, 100]. The composting process involves careful control with nutrient addition, tilling, watering and addition of suitable microbial consortia and bulking materials in the form of organic wastes to improve bioremediation. The composting process requires thermophilic conditions of 50–65°C to properly compost soil contaminated with hazardous compounds such as petroleum hydrocarbon compounds. An increased temperature results from heat generated from the microbial activities during the metabolic breakdown of organic materials in the compost, and efficient degradation of pollutants is achieved by periodic tilling, watering and aeration of the compost [101]. **Figure 12** illustrates the compost piling of contaminated soil.

Atagana [102] conducted composting bioremediation of petroleum hydrocarbons using sewage sludge compost on contaminated soil with a total petroleum hydrocarbon (TPH) concentration of 380,000 mg kg^{-1} for 19 months. The results obtained

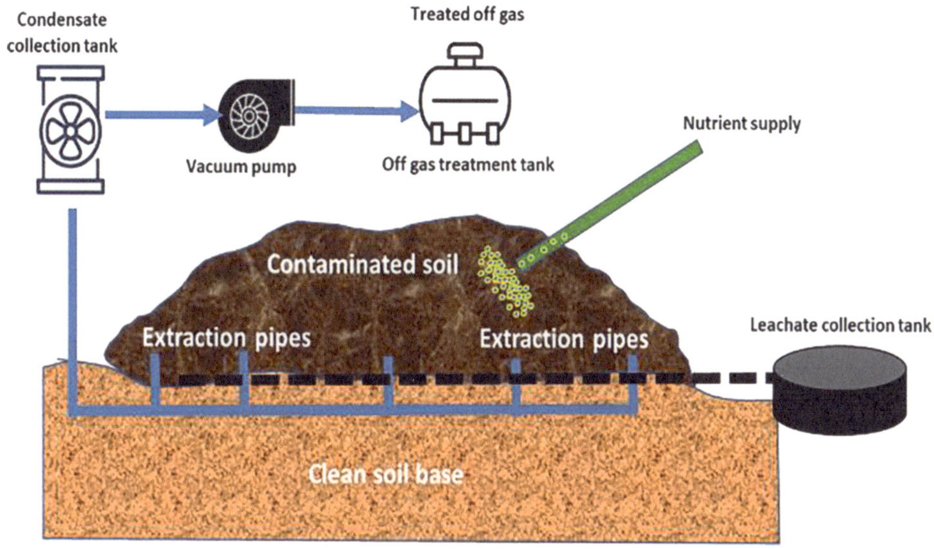

Figure 11.
Biopiling of contaminated soil [94].

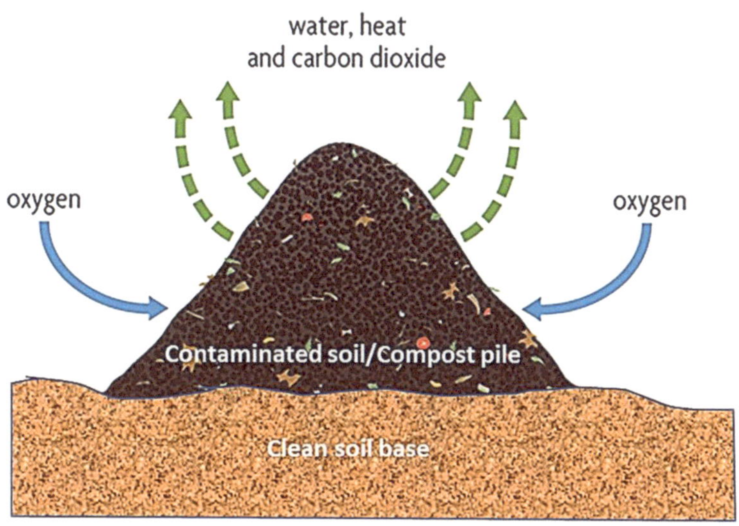

Figure 12.
Contaminated soil composting pile.

after the experiment period showed a 99% removal efficiency for TPH, while other selected hydrocarbon components were removed 100% within the experiment period. Composting helps degrade, bind and convert contaminants into harmless substances and compounds with substantial potential for remediation application to treat petroleum hydrocarbon contaminated soil [103].

The benefits of compost piling include abundant nutrients, soil enrichment retains moisture and nutrients, improves soil quality and altering soil pH, cheap soil conditioner, eco-friendly and cost-effective, and promoting the growth of beneficiary microorganisms. The disadvantages include; it requires extensive monitoring and turning of the pile, takes time and energy, takes about 6 months to

2 years under optimal conditions, emission of greenhouse gases and requirement for a large site area.

2.12 Windrow

The windrow treatment process relies on periodic tilling, ploughing and turning piled contaminated soil with water application to increase moisture and aeration with the distribution of nutrients to enhance biodegradation. In the windrowing process, the increase in microbial activities by the indigenous and transient petroleum hydrocarbon-degrading microorganisms in the contaminated soil speed up the biodegradation process [71, 79]. The biodegradation process is accomplished through biotransformation, assimilation and mineralisation [104]. Compared with biopiling, the windrowing method showed a higher removal efficiency rate for petroleum hydrocarbons. The windrowing process for the remediation of polluted soil is illustrated in **Figure 13**.

A study demonstrated by Al-Daher and Al-Awadhi [105] investigated biodegradation of petroleum hydrocarbon contaminated soil using a windrow soil system for 10 months. The windrow system was subjected to regular watering, tilling and turning to enhance aeration and microbial activities. The results obtained showed a 60% reduction in the total petroleum hydrocarbons (TPH) in the first 8 months, and the degradation rate was enhanced when the moisture content was effectively maintained.

The benefits of the windrowing process include; soil enrichment, retaining moisture and nutrients, improving soil quality and altering soil pH, requiring low capital and operational costs, being eco-friendly and easy to implement and promoting the growth of beneficiary microorganisms. On the downside, windrow treatment is not the best option in removing soil contaminated with volatile petroleum hydrocarbon compounds due to the release of toxic volatile compounds during the periodic turning and tilling [79]. There is an emission of greenhouse gases such as methane (CH_4) in windrow treatment due to the formation of an anaerobic zone within the piled heap [103]. It requires ample space for composting, attracting scavengers, long duration of time under optimal conditions, produces odour, compost may become anaerobic in rainy conditions, requires regular turning to maintain aerobic conditions and vulnerability to climate changes.

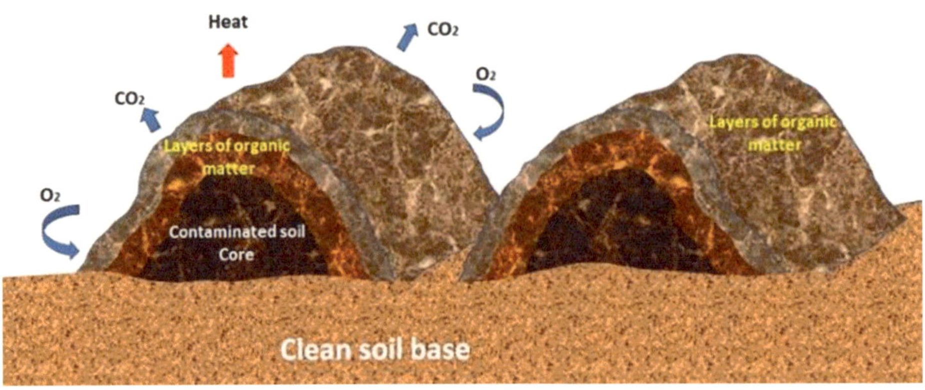

Figure 13.
Windrowing of petroleum hydrocarbon polluted soil.

2.13 Vermiremediation

Vermiremediation is an expanding technology that uses earthworms to biodegrade hazardous contaminated soil [106, 107]. The earthworms in the soil help enhance and improve soil fertility, biological, chemical and physical properties. They stimulate and enhance microbial activities by creating suitable conditions for microorganisms to thrive and improve soil aeration by burrowing and tunnelling through the soil structures [108, 109]. The presence of earthworms in the soil depends on soil moisture, organic matter content and pH. They usually occur in diverse habitats, especially those rich in organic matter and moisture [110, 111]. Vermiremediation of petroleum hydrocarbon in the soil occurs through vermidegradation. The earthworms stimulate the biodegradation processes by enhancing oxidation, soil aeration and microbial activities in the polluted soil. **Figure 14** illustrates the components of vermiremediation in petroleum hydrocarbon contaminated soil.

A study demonstrated by Azizi et al. [113] conducted vermiremediation using earthworm (*Lumbricus rubellus*) to degrade petroleum hydrocarbon components such as polycyclic aromatic hydrocarbons (PAHs), anthracene, phenanthrene and benzo[a]pyrene (BaP) within 30 days. The result obtained showed a removal efficiency of 99.9% for PAHs. Sinha et al. [114] demonstrated a similar study for earthworms remedial action on polycyclic aromatic hydrocarbons (PAHs) contaminated soils in a gasworks site. The result obtained showed 80% removal efficiency for PAHs compared to 21% removal efficiency in microbial degradation.

The benefits of vermiremediation include; minimal environmental disruption, enhanced organic matter, nutrient concentration and biological activity, improved soil utility and fertility, and cost-efficiency. The disadvantages include; high concentration of pollutants may be toxic to the earthworms, the process is restricted to the depth of earthworm activities, effective for slightly or moderately contaminated soil,

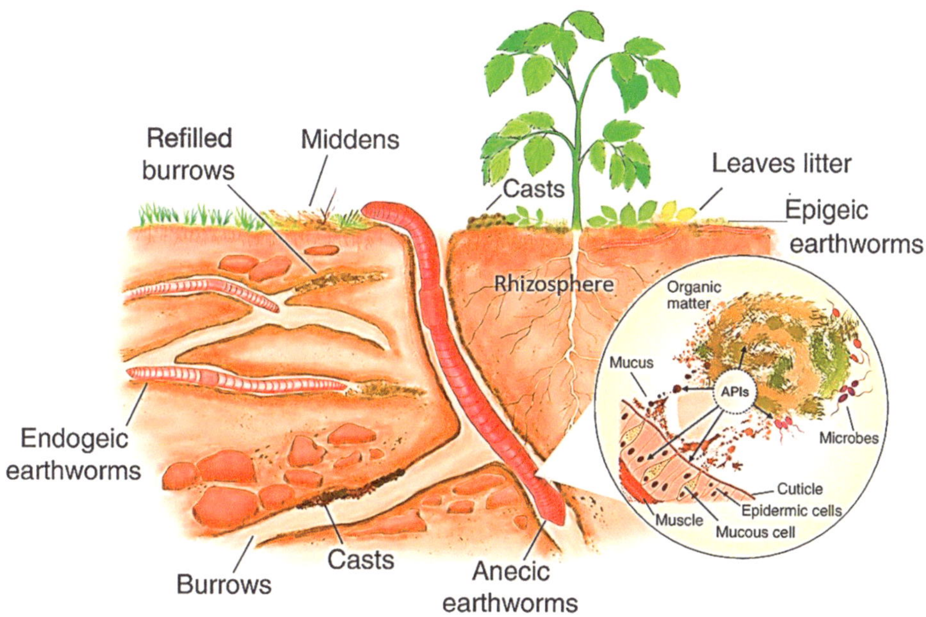

Figure 14.
Vermiremediation in petroleum hydrocarbon contaminated soil [112].

requires strict conditions, sensitive to climate and seasonal conditions, and restricted by food abundance in the soil [106].

2.14 Mycoremediation

Mycoremediation involves using fungi processes to biodegrade hazardous contaminants such as petroleum hydrocarbons to less toxic or non-toxic forms, thereby reducing or eliminating environmental contaminants [115–117]. Fungi can degrade variable environmental recalcitrant pollutants due to their ability to produce and secrete extracellular enzymes such as peroxidases that break down lignin and cellulose [118, 119]. Ligninolytic fungi such as the white-rot fungi *Polyporus* sp. and *Phanaerochaete chrysosporium* are essential in mycoremediation because they can degrade a diverse range of toxic and hazardous pollutants [120]. The degradative action of fungi is effective in various situations where they degrade different materials. When cultivated in polyethene contaminated soil, fungi such as *Penicillium* sp. degrade polyethene effectively [121]. **Figure 15** illustrates the mycoremediation components in petroleum hydrocarbon polluted soil.

Studies have shown that many filamentous fungi species are petroleum hydrocarbon-degrading in nature. Some white rot fungi use their mycelia to degrade petroleum hydrocarbon contaminants due to their high production of oxidative enzymes, extracellular enzymes, chelators and organic acids, which help them degrade petroleum hydrocarbon pollutants [122]. In a mycoremediation study demonstrated by Ulfig et al. [123], keratinolytic fungi *Trichophyton ajelloi* were utilised to remove hexadecane and pristane from crude oil-polluted soil. In another similar study conducted by Njoku et al. [107], *Pleurotus pulmonarius* was used in mycoremediation of soil contaminated with petroleum hydrocarbon mixture comprising petrol, diesel, spent engine oil and spent diesel engine oil lubricant at the ratio of 1:1:1:1 in various concentrations of 2.5%, 5%, 10% and 20% for 62 days period. The results showed that the soil with 10% concentration had a removal efficiency of 68.34% for TPH, while soil with 2.5% concentration yielded 22.12% removal efficiency for TPH.

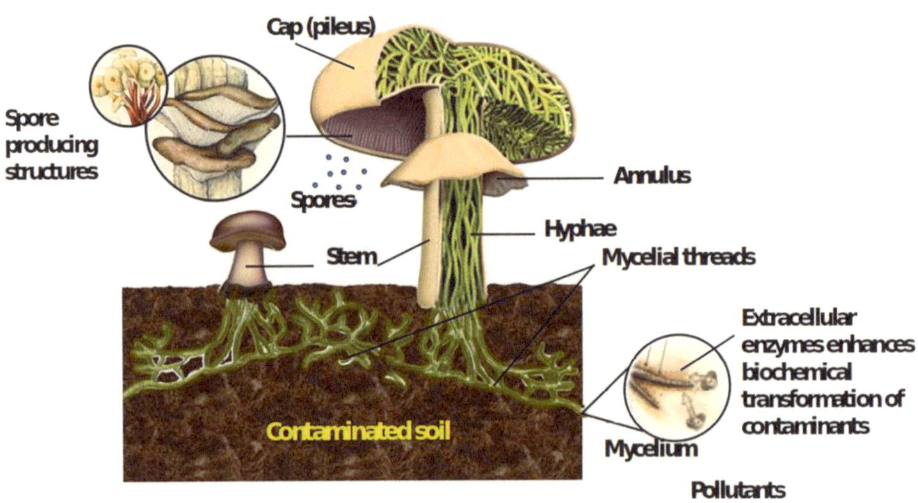

Figure 15.
Mycoremediation of petroleum hydrocarbon polluted soil.

These results suggest that the fungi *Pleurotus pulmonarius* can biodegrade soil contaminated with a moderate level of the petroleum hydrocarbon mixture.

The benefits of mycoremediation include; minimal disturbance to the environment, does not produce corrosive or harmful chemicals, eco-friendly and cost-effective, and requires no special equipment. The disadvantages include; the efficiency is not 100%, long-duration for treatment, periodic turning with reapplication of growth medium is required, competition with indigenous bacterial population may reduce the efficiency, and high concentration of contaminants may be toxic to the fungi.

2.15 Phycoremediation

Phycoremediation, a technique that uses algal species (macroalgae or microalgae) to sequester, remove, break down, biotransform or metabolise pollutants such as petroleum hydrocarbons from contaminated water environments [124–126]. As illustrated in **Figure 16**, this technique is one of the effective methods used in water pollution treatment due to its high efficiency and low-cost usage [127]. Algae can accumulate and degrade toxic pollutants and organic compounds such as petroleum hydrocarbons, biphenyls, pesticides, and phenolics [125]. Algae are very adaptive in most environments and grow in autotrophic, mixotrophic, or heterotrophic conditions. Algae play a vital role in regulating and controlling the concentration of metals in the water environment. The mixotrophic algae are excellent in bioremediation and carbon sequestration [128].

Algae can produce O_2, fix CO_2 by photosynthetic process, increase the BOD level in the polluted water, and remove excess nutrients [129]. The mineral uptake by microalgae occurs in two steps. The initial step is independent of cell processes and involves physical adsorption onto the cell's surface, and the ions are gradually carried into the cell by chemisorption [120]. The second step is dependent on cell processes and involves intracellular uptake and absorption. Studies have shown that heavy metals can be sequestered in the polyphosphate body of algae and serve for detoxification and storage [130]. Phycoremediation was successfully used to reduce nutrient levels in wastewater treatment, and the technique includes algal biofilm, algal turf scrubbers,

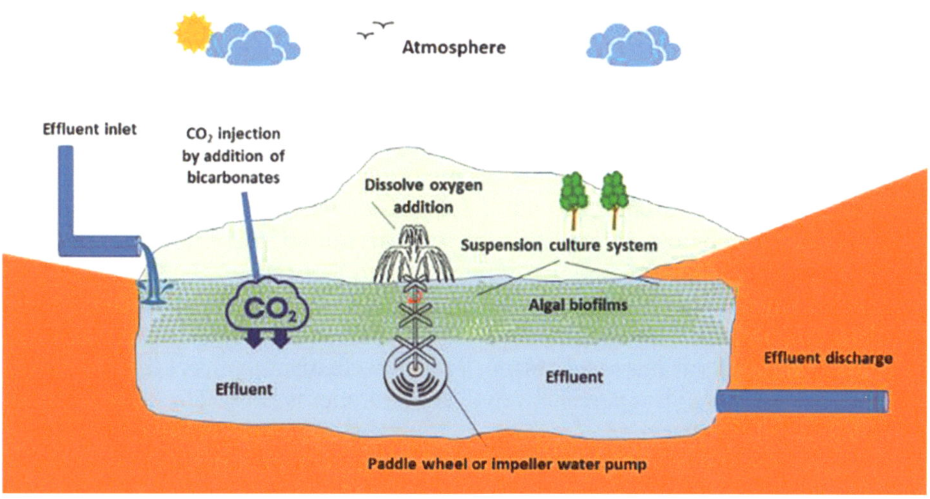

Figure 16.
Phycoremediation technique in a pond system.

high-rate algal ponds, and immobilised algae [127]. Several algae species such as *Chlamydomonas*, *Chlorella*, *Botryococcus* and *Phormidium* are involved in phycoremediation. The use of microalgae in the phycoremediation of petroleum hydrocarbon is gaining interest as some algae species can degrade and oxidise hazardous petroleum hydrocarbon components into less noxious compounds [131, 132].

A phycoremediation study was demonstrated by Kalhor et al. [133], who investigated the potential of *Chlorella vulgaris* in biodegradation of the crude oil-contaminated water environment. Different crude oil concentrations were prepared and treated in their investigation, and the removal efficiency was calculated after the incubation period. The result obtained after 14 days incubation period showed that aromatic hydrocarbon compounds (benzene and naphthalene) and alkane (nonadecane) were biodegraded at the removal efficiencies of 89.17% at 10 g/l and 76.53% at 20 g/l concentration by the algae. Their result confirmed that the algae *C. vulgaris* could remove light components of petroleum hydrocarbon compounds in the contaminated water.

The advantages of phycoremediation include; simple and economic pilot scale, low implementation cost, high versatility and adaptability, high nutrient removal in effluents, algal biomass is easy and cheap to harvest in low scale operation, and the algal biomass can be used for biogas production. The disadvantages include; it is difficult and expensive to harvest algal biomass in large scale operations, poor and inconsistent contaminant removal due to characteristics of the pollutants, sensitivity to climate and seasonal conditions, the infestation of predators that feed on algae, and injection of CO_2 incur a cost for the implementation.

2.16 Phytoremediation

Phytoremediation is a low-cost remediation technique that uses green plants and the associated soil microorganisms to reduce the concentrations of contaminants and their toxic effects [134]. The technique removes, extracts, and sequesters the contaminants (decontamination) into the plant matrix (stabilisation) [43]. Phytoremediation uses the natural processes of the green plants or plant-based systems to remediate environments contaminated by organic compounds, heavy metals, and inorganic compounds. It formed the basis of the reed beds and constructed wetlands [43]. The phytoremediation system uses the synergistic relationship among the plants, indigenous microorganisms dwelling in the contaminated soil, and the roots of the plants [135]. The plants produce inherent enzymatic activities and uptake processes that remove and sequester contaminants. The plants act as symbiotic hosts to aerobic and anaerobic microorganisms, providing nutrients and habitat to the microorganisms [134]. The mechanisms of phytoremediation include phytoextraction (phytoaccumulation), phytodegradation, phytostabilisation, phytotransformation, phytovolatilisation, rhizofiltration, and rhizodegradation (rhizoremediation), as illustrated in **Figure 17** [137, 138].

In phytoremediation, plants break down, degrade, concentrate, sequester, bioaccumulate, contain, stabilise and metabolise contaminants by acting as filters or traps in the tissue through various mechanisms. These mechanisms convert the contaminants into less toxic and less persistent in the environments [139]. The mechanisms and efficiency of the phytoremediation technique depend on the pollutants, bioavailability, and properties of the polluted soil, and the mechanisms affect the mobility, toxicity of pollutants, volume, and concentration [136, 140]. The plants' roots and shoots provide colonisable surface area for absorption, exudates, and leachates in the rhizosphere for microbial activities [141]. The success of phytoremediation depends

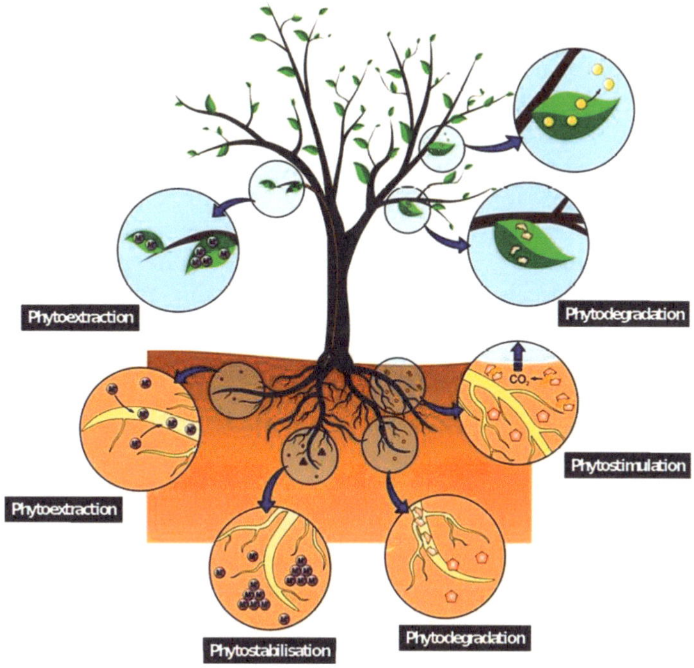

Figure 17.
Mechanism of phytoremediation [136].

mainly on the plant's ability to bioassimilate or bioaccumulate both organic and inorganic contaminants into their cell wall structures and carry out oxidative degradation of organic xenobiotics [142].

Many researchers have conducted phytoremediation and reported studies using different plants to remediate soil contaminated with petroleum hydrocarbons, heavy metals and other organic pollutants. Cook and Hesterberg [143] published a summary of major plants (trees and grasses) currently used in phytoremediation, which adsorb or degrade contaminants in polluted environments. Other researchers, including Dadrasnia and Agamuthu [144], Cartmill et al. [145] and Agamuthu et al. [146], demonstrated phytoremediation of petroleum hydrocarbon contaminated soil using several plants with the addition of organic wastes and organic fertilisers to enhance the biodegradation process.

Some of the advantages of phytoremediation include; it is a permanent treatment technique, it has low capital investment and operation costs, there is no soil excavation, phyto-accumulated metals may be recycled and provides additional economic advantages, it eliminates secondary air and water-borne wastes, and it has public acceptance due to aesthetic reasons. The disadvantages include being slower than other remediation techniques, hyperaccumulating plants being slow growers, working efficiency is not 100%, may not be effective for mixture pollutants, high concentration of contaminants may be toxic to plants, and treatment is limited to shallow contaminants.

2.17 Electrobioremediation

Electrobioremediation or bioelectrochemical system is an emerging biodegradation technology with a trans-disciplinary system that depends on the use of electroactive microorganisms to catalyse the oxidation or reduction reactions of organic and inorganic electron donors. The bioelectrochemical system delivers electrons to the solid-state electrode (anode), with subsequent transfer or exchange of electrons to the solid-state electrode (cathode) through a conductive circuit and simultaneously generating electrical energy (**Figure 18**) [147, 148]. The mechanism involves an electrokinetic process in the acceleration and orientation of the transport of pollutants and microorganisms [149].

Bioelectrochemical system works effectively in contaminated media as unlimited electron acceptors or donors [150] and converts chemical energy from organic wastes or contaminants to electrical energy and hydrogen or value-added chemical products [151]. The system works on the interface of electrochemistry and fermentation [152]. The bioelectrochemical system can be classified based upon the application of microbial fuel cells for power generation, microbial electrolytic cells for biofuel production, microbial desalination cell for saline water desalination, and microbial electro synthetic cells for the synthesis of value-added by-products [134].

A study conducted by Daghio et al. [77] demonstrated that bioelectrochemical systems energised and stimulated anaerobic oxidation of different types of organic wastes to reduce contaminants in soil and groundwater, including petroleum hydrocarbons halogenated compounds. In a laboratory study, Palma et al. [153] demonstrated a bioelectrochemical treatment system for petroleum hydrocarbon contaminated groundwater. The results showed that phenols were gradually removed from 12 to 50% while electric current generation gradually increased from 0.3 mA to 1.9 mA. The phenol removal rate and the coulombic efficiencies were 23 ± 1 mg L^{-1} d and 72 ± 8% on average.

The advantages of electrobioremediation include generating electrical energy level and electron flux; no waste is generated, cheap operational cost, and highly selective

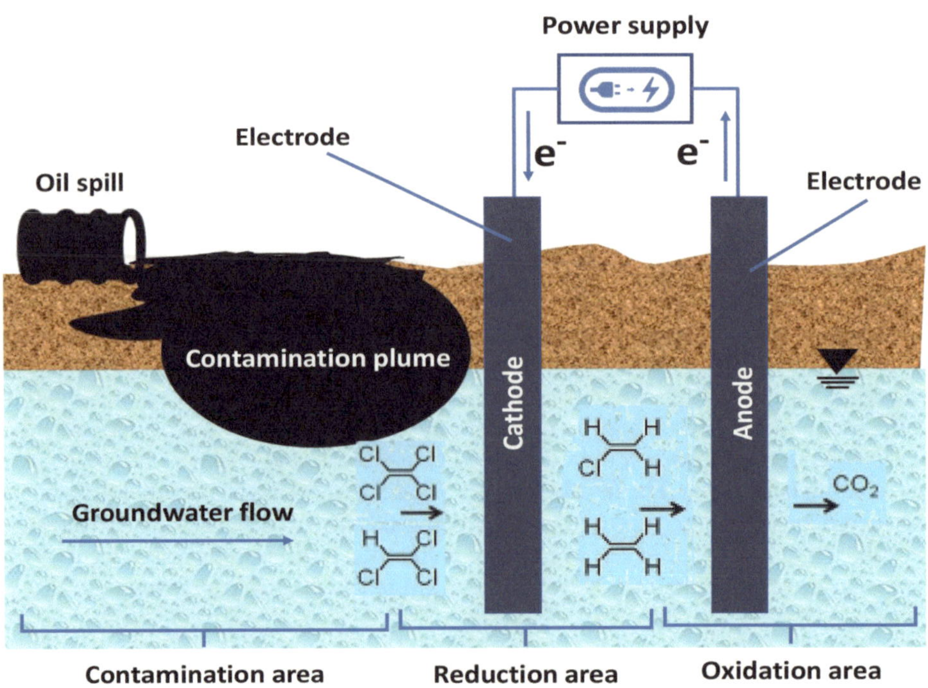

Figure 18.
In situ *electrobioremediation of oil-polluted soil [77].*

towards target pollutants, pollutants can be adsorbed on the electrodes when graphite or carbon is used. The disadvantages include; slower anaerobic degradation than aerobic degradation. The cathodic reaction may limit the anodic reaction when microbial fuel cells are used, chlorine gas is produced, a scale-up process is challenging, and the process is affected by changes in pH in the contaminated soil [77].

2.18 Nanobioremediation

Nanobioremediation is an emerging technology used in remediating environmental pollutions. The system functions with the aid of reactive biosynthetic nanomaterials (NMs), nanoparticles (NPs), nanostructured materials (NSMs), nanocomposites manufactured particles (NCMPs), manufactured nanoparticles (MNPs), and nanoclusters (NCs) [154–156]. These biosynthetic nanoparticles exhibit unique physical, chemical and biochemical properties in enzyme-mediated remediation, transformation, and detoxification of persistent hydrophobic contaminants and toxicants [157]. These nanomaterials or particles are engineered or formed by plants or microorganisms and comprise particles with at least one dimension measuring between 1.0 and 100 nm [158, 159]. **Figure 19** illustrates *in situ* nanobioremediation of oil-polluted soil.

The nanoparticles can be carbon-based (carbon fullerenes) and carbon nanotubes. They can be metal-based (quantum dots, nano zero-valent iron (nZVI), nanosilver, nanogold, and nanosized metal oxides such as ZnO, Fe_3O_4, TiO_2, CeO_2). They can also be dendrimers or nano polymers and composite or bulk-type materials [161]. The nanomaterial or nanoparticles have properties that allow catalysis and chemical reduction to remove the contaminants. As reducing agents, the particles degrade hazardous organic contaminants in the environment. The process changes elements' oxidation state, combined with catalytic enhancement of redox reactions for soil and groundwater remediation.

In the nanoremediation process, no groundwater is pumped out for above-ground treatment, and no soil is excavated or transported to a different location for disposal

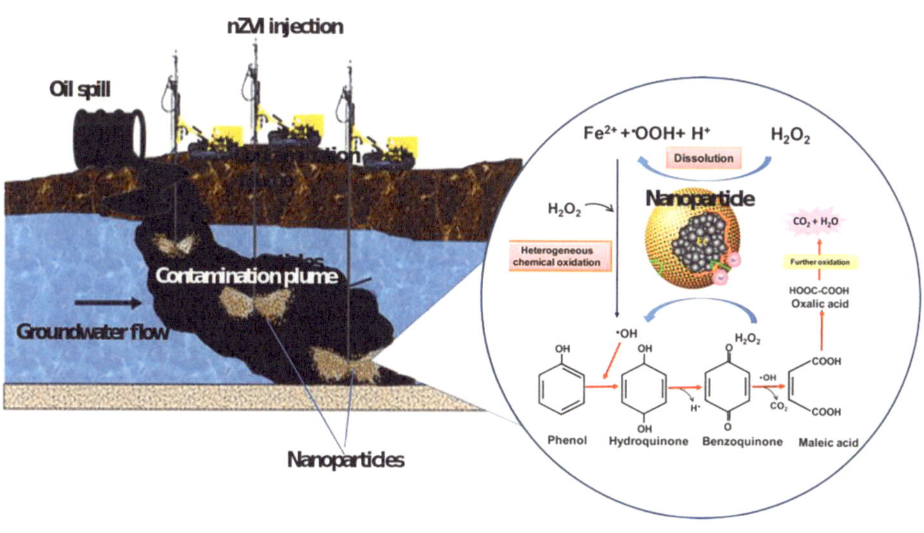

Figure 19.
In situ *nanoremediation in oil-polluted soil [160].*

and treatment [162]. With the nanoparticles' minute size and innovative surface coating, they pervade tiny spaces in the subsurface and remain dispersed in the soil or groundwater, allowing the particles to move and migrate farther than larger or micro or macro-sized particles and achieve wider distribution [163]. The sorption process occurs by adsorption and absorption. In adsorption, the interactions between the pollutants and the sorbent occur at the surface level, while in absorption, the pollutants penetrate deeper into the sorbent layers to form a solution [164]. The mobility of natural or biosynthetic nanoparticles depends on their dispersions, aggregations, settlings, and formation of mobile clusters.

Nanoparticles such as zeolites, carbon nanotubes, nanofibres, metal oxides, titanium dioxide, enzymes, and noble metals such as bimetallic nanoparticles (BNPs) have been used successfully in the remediation of organic compounds and petroleum hydrocarbons from the contaminated environments [165, 166]. Among the nanoparticles, the most widely used is the nanoscale zero-valent iron (nZVI) modified with palladium inclusion as a catalyst for improved performance [167]. Nanobioremediation can be used where other conventional remediation technologies do not prove productive because nanoparticles are less toxic to soil flora and enhance microbial activity [157]. The nanoparticles have highly desired properties for *in situ* applications due to the nanosize and innovative surface coatings. The particles easily penetrate tiny spaces in the subsurface, remain suspended in groundwater, and allow further migration and wider distribution [163].

A study conducted by Reddy et al. [168] demonstrated nanobioremediation using nanoscale iron to degrade the organic compound dinitrotoluene (DNT) in the soil. The results obtained showed 41–65% removal efficiency for DNT near the anode, while removal efficiency of 30–34% was recorded near the cathode. The highest removal was recorded using lactate-modified nanoscale iron particles. However, the overall degradation of DNT was due to nanoscale iron particles having the electrochemical process that enhanced the delivery of nanoscale particles in the degradation of organic contaminants.

The advantages of nanobioremediation include; effectivity across a wide range of environmental conditions, the high surface area increasing reactivity and treatability, extending the range of treatable contaminants, eliminating intermediate by-products, and combining with other treatment techniques for enhanced remediation. The disadvantages include; potential to generate harmful by-products, the potential to enter the food chain with the possibility of biomagnification and bioaccumulation, the production of nanoparticles is an expensive engineering process, and the societal issue due to fear of the environmental impact from the manufactured nanoparticles.

2.19 Trichoremediation

Trichoremediation is an emerging technique. The etymology originates from the ancient Greek word θρίξ *(tricho),* meaning "hair," and Latin word (*remedium*), meaning "restoring balance." It describes a biological treatment of environmental contaminants by utilising hairs (keratinaceous materials) to increase the metabolic activities of the keratinolytic and keratinophilic microbes with pollutant degrading abilities in the co-metabolic degradation of the substrates [134]. The microorganisms display lipolytic activity and remove petroleum hydrocarbons from the medium during biodegradation [123, 169]. Trichoremediation involves biostimulation of indigenous microorganisms in the contaminated soil and bioaugmentation with the naturally associated microorganisms inhabiting the hair materials. Additional mechanisms that

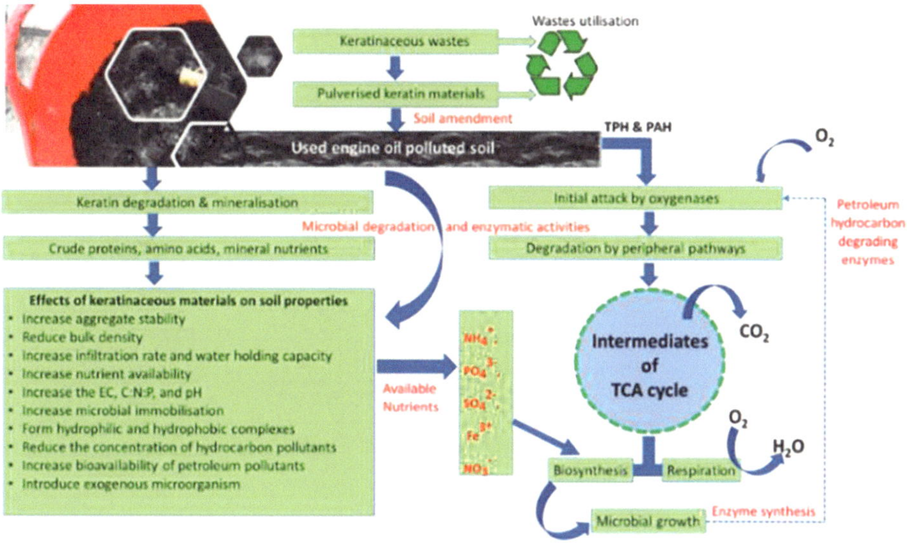

Figure 20.
Trichoremediation of petroleum hydrocarbon polluted soil.

participate in the process are absorption and adsorption due to the chemisorption properties of hairs [170–172]. **Figure 20** illustrates the components of trichoremediation for petroleum hydrocarbon contaminated soil.

Cervantes-González et al. [173] investigated the ability of chicken feather wastes as petroleum hydrocarbon sorbent and studied their structural biodegradation and removal of petroleum hydrocarbons. Their findings showed that chicken feathers enhanced the contact between petroleum hydrocarbons and bacteria and enhanced the removal of petroleum hydrocarbons. They also observed that the microorganisms colonised the chicken feathers and degraded the materials completed in the study. In their observation during the treatment, there was an exponential growth phase of bacteria during the early days of the treatment, and the simultaneous degradation of feathers and petroleum hydrocarbons was evident [173].

The benefits of trichoremediation technology include; relatively low cost and maintenance, ease of implementation and operation, reduced landfill wastes, fully organic and biodegradable materials, improved soil quality and structure, and additional accessible carbon sources and co-metabolites. The disadvantages include; long treatment time, sensitivity to the level of toxicity and environmental conditions, generating toxic metabolites, metabolic pathways may switch to a less toxic carbon source, inhibits metabolic pathway by the presence of the metabolites, and additional compounds may negatively affect the biodegradation process.

3. Factors affecting the biological treatment technologies

The purpose of biological treatment technologies for biodegradation of petroleum hydrocarbon polluted sites through sustainable and eco-friendly means is to eliminate the hazards of pollution in the environment and human health risks. Applying biological treatment technology in a polluted environment at a field scale is a challenging and laborious task. The choice of a biological treatment technology

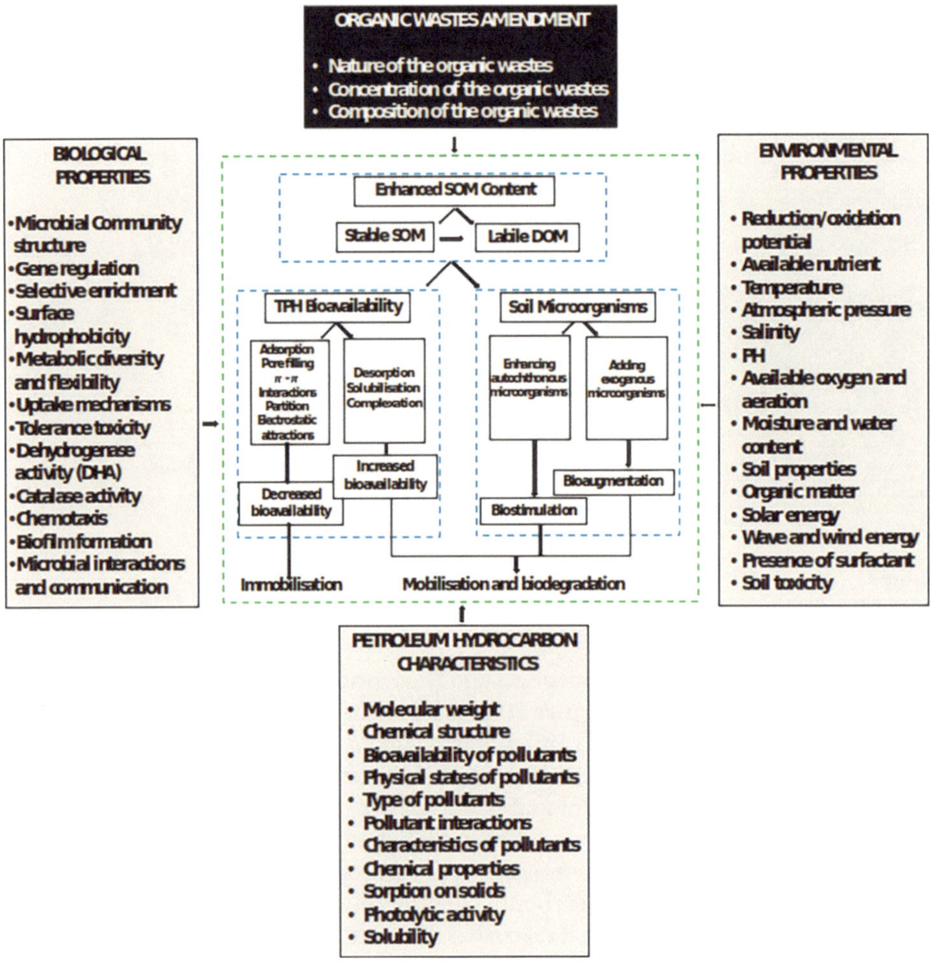

Figure 21.
Factors affecting the degradation of petroleum hydrocarbons polluted using organic wastes amendments [134].

depends on several biological and environmental properties, which vary from one site to another. The influencing parameters comprised environmental and biological properties include nature and concentration of the contaminants, type and properties of the soil, and the interaction with microorganisms and metabolic pathways [174]. The environmental properties influence the biological properties, while the biological properties produce the overall biodegradation effect in the system. The environmental properties affecting biodegradation influence the rates and extent of microbial transformation of the pollutants [175]. Biological treatment technologies immobilise contaminants through adsorption, absorption, desorption, volatilisation, solubilisation, complexation, hydrolysis, oxidation, and mineralisation [12, 13]. **Figure 21** illustrates the various factors affecting biological treatment technologies.

4. Conclusions

The biological treatment technologies have grown as alternatives to the traditional physicochemical, thermal and electromagnetic technologies for the remediation of petroleum hydrocarbons polluted soil. They are preferred due to low energy consumption, cost-effectiveness, environmental-friendliness, non-invasiveness, feasibility, and sustainability compared to other physicochemical, thermal and electromagnetic treatment options, which are cost-prohibitive, often destroy the soil properties and render the soil impoverished and sterile eventually. The biological treatment technologies can be selectively adapted and adopted to degrade the pollutants without causing further damage to the site and the indigenous flora and fauna. Although various biological treatment technologies are accessible, no single biological treatment is the most suitable for all varieties of contaminants and the type of site-specific conditions occurring in the petroleum hydrocarbon-affected environments. Good knowledge of the environmental conditions of the affected environments, nature, composition and properties of the contaminants, fate, transport, and distribution of the contaminants, mechanism of biodegradation, the interactions and relationships with the microorganisms, intrinsic and extrinsic factors affecting the remediation processes, and the potential impact of the possible remedial measure determine the choice and selection of a biological treatment technology requirements. More than one biological treatment technology may be adopted or combined into a process train to effectively remove, contain or destroy the petroleum hydrocarbon pollutants in polluted environments.

However, selecting one or more biological treatment technology is essential in decision-making, as many parameters that conflict in nature plays a significant role in decision-making. Consequently, it is a welcome idea to select biological treatment technologies that are feasible, adaptive, scientifically defensible, sustainable, non-invasive, eco-friendly, and economical because remediation of petroleum hydrocarbon polluted environments through the conventional physicochemical, thermal, and electromagnetic technologies is a challenging, laborious, extensive and expensive task.

Acknowledgements

The authors express sincere gratitude to all the researchers whose valuable data reported in their respective publications were cited in this chapter and contributed to the knowledge of the biological treatment technologies. We are also grateful to the reviewers whose constructive criticisms have benefitted the manuscript and brought it to the present form.

Conflict of interest

The authors hereby declare that there is no conflict of interest regarding the publication of this book chapter.

Author details

Innocent Chukwunonso Ossai[1,2*], Fauziah Shahul Hamid[1,2] and Auwalu Hassan[1,2,3]

1 Faculty of Science, Institute of Biological Sciences, University of Malaya, Kuala Lumpur, Malaysia

2 Faculty of Science, Centre for Research in Waste Management, University of Malaya, Kuala Lumpur, Malaysia

3 Faculty of Science, Department of Biological Sciences, Federal University Kashere, Gombe State, Nigeria

*Address all correspondence to: ossainonso@yahoo.com

References

[1] Ngene S, Tota-Maharaj K, Eke P, Hills C. Environmental and economic impacts of crude oil and natural gas production in developing countries. International Journal of Energy, Environment, and Economics. 2016;**1**(3):64-73

[2] Inkpen A, Moffett MH. The Global Oil and Gas Industry: Management, Strategy and Finance. Tennessee, USA: PennWell Books LLC; 2011

[3] Jafarinejad S. Introduction to the petroleum industry. In: Petroleum Waste Treatment and Pollution Control. Oxford, UK: Butterworth-Heinemann; 2017. pp. 1-17

[4] Fanchi JR, Christiansen RL. Introduction to Petroleum Engineering. Hoboken, New Jersey. USA: John Wiley and Sons Inc.; 2017

[5] Mariano J, La Rovere E. Environmental Impacts of the Oil Industry. Saarbrücken, Germany: LAP Lambert Academic Publishing; 2017

[6] Sajna KV, Sukumaran RK, Gottumukkala LD, Pandey A. Crude oil biodegradation aided by biosurfactants from *Pseudozyma* sp. NII 08165 or its culture broth. Bioresource Technology. 2015;**191**:133

[7] Souza EC, Vessoni-Penna TC, Oliveira RPD-S. Biosurfactant-enhanced hydrocarbon bioremediation: An overview. International Biodeterioration and Biodegradation. 2014;**89**:88-94

[8] Bejarano AC, Michel J. Large-scale risk assessment of polycyclic aromatic hydrocarbons in shoreline sediments from Saudi Arabia: Environmental legacy after twelve years of the Gulf war oil spill. Environmental Pollution. 2010;**158**(5):1561-1569

[9] Brown DM, Bonte M, Gill R, Dawick J, Boogaard PJ. Heavy hydrocarbon fate and transport in the environment. Quarterly Journal of Engineering Geology & Hydrogeology. 2017;**50**:333-346

[10] Hreniuc M, Coman M, Cioruța B. Consideration regarding the soil pollution with oil products in Sacel-Maramures. In: International Conference of Scientific Research and Education in the Air Force (AFASES). Brasov, Romania; 2015. pp. 557-562

[11] Truskewycz A, Gundry TD, Khudur LS, Kolobaric A, Taha M, Aburto-Medina A, et al. Petroleum hydrocarbon contamination in terrestrial ecosystems—Fate and microbial responses. Molecules. 2019;**24**(18):3400

[12] Brassington KJ, Pollard SJT, Coulon F. Weathered hydrocarbon biotransformation: Implications for bioremediation, analysis and risk assessment. In: Temmis KN, editor. Handbook of Hydrocarbon and Lipid Microbiology. Berlin, Heidelberg: Springer; 2010

[13] Abdel-Shafy HI, Mansour MS. A review on polycyclic aromatic hydrocarbons: Source, environmental impact, effect on human health and remediation. Egyptian Journal of Petroleum. 2016;**25**:107-123

[14] Balseiro-Romero M, Monterroso C, Casares JJ. Environmental fate of petroleum hydrocarbons in soil: Review of multiphase transport, mass transfer, and natural attenuation processes. Pedosphere. 2018;**28**:833-847

[15] Esbaugh AJ, Mager EM, Stieglitz JD, Hoenig R, Brown TL, French BL, et al. The effects of weathering and chemical dispersion on Deepwater horizon crude oil toxicity to mahi-mahi (*Coryphaena hippurus*) early life stages. Science of the Total Environment. 2016;**543**:644-651

[16] Mishra AK, Kumar GS. Weathering of oil spill: Modeling and analysis. Aquatic Procedia. 2015;**4**:435-442

[17] Nzila A, Razzak SA, Zhu J. Bioaugmentation: An emerging strategy of industrial wastewater treatment for reuse and discharge. International Journal of Environmental Research and Public Health. 2016;**13**:846

[18] Mckee RH, Adenuga MD, Carillo J-C. Characterisation of the toxicological hazards of hydrocarbon solvents. Critical Reviews in Toxicology. 2015;**45**(4):273-365

[19] Maletic SP, Dalmacija BD, Roncevic SD. Petroleum hydrocarbon biodegradability in soil—Implications for bioremediation. In: Hydrocarbon. Vol. 3. London, UK: IntechOpen; 2013. pp. 43-64

[20] Fine P, Graber E, Yaron B. Soil interactions with petroleum hydrocarbons: Abiotic processes. Soil Technology. 1997;**10**:133-153

[21] Kristensen AH, Henriksen K, Mortensen L, Scow KM, Moldrup P. Soil physical constraints on intrinsic biodegradation of petroleum vapours in a layered subsurface. Vadose Zone Journal. 2010;**9**(1):137-147

[22] Hamzah A, Chia-Wei P, Pek-Hoon Y, Nurul H. Oil palm empty fruit bunch and sugarcane bagasse enhance the bioremediation of soil artificially polluted by crude oil. Soil and Sediment Contamination: An International Journal. 2014;**23**(7):751-762

[23] Lim MW, Lau EV, Poh PE. A comprehensive guide of remediation technologies for oil contaminated soil—Present works and future directions. Marine Pollution Bulletin. 2016;**109**:14-45

[24] Margesin R, Labbe D, Schinner F, Greer CW, Whyte LG. Characterisation of hydrocarbon-degrading microbial populations in contaminated and pristine alpine soils. Applied and Environmental Microbiology. 2003;**69**(6):3085-3092

[25] Yanti MD. Bioremediation of petroleum-contaminated soil: A review. IOP Conference Series: Earth and Environmental Science. 2018;**118**:1-7

[26] Dzionek A, Wojcieszyńska D, Guzik U. Natural carriers in bioremediation: A review. Electronic Journal of Biotechnology. 2016;**23**:28-36

[27] Rughöft S, Jehmlich N, Gutierrez T, Kleindienst S. Comparative proteomics of *Marinobacter* sp. TT1 reveals corexit impacts on hydrocarbon metabolism, chemotactic motility, and biofilm formation. Microorganisms. 2021;**9**(1):3

[28] Machando LF, Leite DCD-A, Rachid CTCD-C, Paes JE, Martins EF, Peixoto RS, et al. Tracking mangrove oil bioremediation approaches and bacterial diversity at different depths in an in situ mesocosms system. Frontiers in Microbiology. 2019;**10**:2107

[29] Wu M, Li W, Dick WA, Ye X, Chen K, Kost D, et al. Bioremediation of hydrocarbon degradation in a petroleum-contaminated soil and microbial population and activity determination. Chemosphere. 2017;**169**:124-130

[30] Xu X, Liu W, Tian S, Wang W, Qi Q, Jiang P, et al. Petroleum hydrocarbon-degrading bacteria for the remediation of oil pollution under aerobic conditions: A perspective analysis. Frontiers in Microbiology. 2018;**9**:2885

[31] Santos JCO. Recovery of used lubricating oils—A brief review. Progress in Petrochemical Science. 2018;**1**(4):000516

[32] Wang ZY, Xu Y, Zhao J, Li FM, Gao DM, Xing BS. Remediation of petroleum contaminated soils through composting and rhizosphere degradation. Journal of Hazardous Materials. 2011;**190**:677-685

[33] Wongsa P, Tanaka M, Ueno A, Hasanuzzaman M, Yumoto I, Okuyama H. Isolation and characterisation of novel strains of *Pseudomonas aeruginosa* and *Serratia marcescens* possessing high efficiency to degrade gasoline, kerosene, diesel oil, and lubricating oil. Current Microbiology. 2004;**49**:415-422

[34] Mancera-López ME, Esparza-García F, Chávez-Gómez B, Rodríguez-Vázquez R, Saucedo-Castañeda G, Barrera-Cortés J. Bioremediation of an aged hydrocarbon-contaminated soil by a combined system of biostimulation–bioaugmentation with filamentous fungi. International Biodeterioration and Biodegradation. 2008;**61**:151-160

[35] Potin O, Rafin C, Veignie E. Bioremediation of an aged polycyclic aromatic hydrocarbons (PAHs)-contaminated soil by filamentous fungi isolated from the soil. International Biodeterioration and Biodegradation. 2004;**54**:45-52

[36] Hoff RZ. Bioremediation: An overview of its development and use for oil spill cleanup. Marine Pollution Bulletin. 1993;**26**:476-481

[37] Bragg JR, Prince RC, Harner EJ, Atlas RM. Effectiveness of bioremediation for the Exxon Valdez oil spill. Nature. 1994;**368**:413-418

[38] Atlas RM. Microbial degradation of petroleum hydrocarbons: An environmental perspective. Microbiological Reviews. 1981; **45**(1):180-209

[39] Duke NC. Oil spill impacts on mangroves: Recommendations for operational planning and action based on a global review. Marine Pollution Bulletin. 2016; **109**:700-715

[40] Iosob G-A, Prisecaru M, Stoica I, Călin M, Cristea TO. Biological Remediation of Soil Polluted with Oil Products: An Overview of Available Technologies. Bacău, Romania: Studii şi Cercetări, Biologie; 2016. pp. 89-101

[41] Wu M, Dick WA, Li W, Wang X, Yang Q, Wang T, et al. Bioaugmentation and biostimulation of hydrocarbon degradation and the microbial community in a petroleum contaminated soil. International Biodeterioration and Biodegradation. 2016;**107**:158-164

[42] Agarry S, Latinwo GK. Biodegradation of diesel oil in soil and its enhancement by application of bioventing and amendment with brewery waste effluents as biostimulation-bioaugmentation agents. Journal of Ecological Engineering. 2015;**16**(2):82-91

[43] Adam G. A study into the potential of phytoremediation for diesel fuel contaminated soil. environmental, agricultural and analytical chemistry [doctoral thesis]. Scotland: University of Glasgow; 2001

[44] Desay C, Pathak H, Madamwar D. Advances in molecular and "-omics"

technologies to gauge microbial communities and bioremediation at xenobiotic/anthropogeny contaminated sites. Bioresource Technology. 2010; **101**:1558-1569

[45] Hazen TC. Biostimulation. In: Timmis KN, editor. Handbook of Hydrocarbon and Lipid Microbiology. Berlin Heidelberg: Springer-Verlag; 2010

[46] Wu M, Wu J, Zhang X, Ye X. Effect of bioaugmentation and biostimulation on hydrocarbon degradation and microbial community composition in petroleum-contaminated loessal soil. Chemosphere. 2019;**237**:124456

[47] Singh B, Bhattacharya A, Channashettar VA, Jeyaseelan CP, Gupta S, Sarma PM, et al. Biodegradation of oil spill by petroleum refineries using consortia of novel bacterial strains. Bulletin of Environmental Contamination and Toxicology. 2012;**89**:257-262

[48] Goswami M, Chakraborty P, Mukherjee K, Mitra G, Bhattacharyya P, Dey S, et al. Bioaugmentation and biostimulation: A potential strategy for environmental remediation. Journal of Microbiology & Experimentation. 2018;**6**(5):223-231

[49] Simpanen S, Dahl M, Gerlach M, Mikkonen A, Malk V, Mikola J, et al. Biostimulation proved to be the most efficient method in the comparison of in situ soil remediation treatments after a simulated oil spill accident. Environmental Science and Pollution Research. 2016;**23**:25024-25038

[50] Agamuthi P, Tan YS, Fauziah SH. Bioremediation of hydrocarbon contaminated soil using selected organic wastes. Procedia Environmental Sciences. 2013;**18**:694-702

[51] Yousefi K, Mohebbi A, Pichtel J. Biodegradation of weathered petroleum hydrocarbons using organic waste amendments. Applied and Environmental Soil Science. 2021:1-12. Article ID 6620294

[52] Zhang C, Wu D, Ren H. Bioremediation of oil-contaminated soil using agricultural wastes via microbial consortium. Scientific Reports. 2020;**10**:9188

[53] Nwankwegu AS, Onwosi CO. Bioremediation of gasoline contaminated agricultural soil by bioaugmentation. Environmental Technology and Innovation. 2017;**7**:1-11

[54] Poi G, Aburto-Medina A, Mok PC, Ball AS, Shahsavari E. Large-scale bioaugmentation of soil contaminated with petroleum hydrocarbons using a mixed microbial consortium. Ecological Engineering. 2017;**102**:64-71

[55] Kästner M, Miltner A. Application of compost for effective bioremediation of organic contaminants and pollutants in soil. Applied Microbiology and Biotechnology. 2016;**100**:3433-3449

[56] Alexandar M. Aging, bioavailability and overestimation of risk from environmental pollutants. Environmental Science & Technology. 2000;**34**(20):4259-4265

[57] Abatenh E, Gizaw B, Tsegaye Z, Wassie M. Application of microorganisms in bioremediation—Review. Journal of Environmental Microbiology. 2017;**1**(1):2-9

[58] Varjani SJ, Rana DP, Jain AK, Bateja S, Upasani VN. Synergistic *ex situ* biodegradation of crude oil by halotolerant bacterial consortium of indigenous strains isolated from onshore sites of Gujarat, India. International Biodeterioration and Biodegradation. 2015;**103**:116-124

[59] Covino S, D'Annibale A, Stazi SR, Cajthami T, Cvancarova M, Stella T, et al. Assessment of degradation potential of aliphatic hydrocarbons by autochthonous filamentous fungi from a historically polluted clay soil. Science of the Total Environment. 2015;**505**:545-554

[60] Raper E, Stephenson T, Anderson DR, Fisher R, Soares A. Industrial wastewater treatment through bioaugmentation. Process Safety and Environmental Protection. 2018;**18**:178-187

[61] Vásquez-Murrieta MS, Hernández-Hernández OJ, Cruz-Maya JA, Cancino-DíazJC,Jan-RobleroJ.Approaches for removal of PAHs in soils: Bioaugmentation, biostimulation and bioattenuation. In: Environmental Sciences. Soil Contamination—Current Consequences and Further Solutions. London, UK: IntechOpen; 2016

[62] Chapelle F, Widdowson H, Brauner JS, Mendez E III, Cassey CC. Methodology for estimating times of remediation associated with monitored natural attenuation. US Geological Survey. Water-Resources Investigations Reports 03-4057. 2003

[63] Trulli E, Morosini C, Rada EC, Torretta V. Remediation in situ of hydrocarbons by combined treatment in a contaminated alluvial soil due to an accidental spill of LNAPL. Sustainability. 2016;**8**:1086

[64] Camenzuli D, Freidman BL. On-site and in situ remediation technologies applicable to petroleum hydrocarbon contaminated sites in the Antarctic and artic. Polar Research. 2015;**34**:24492

[65] Iturbe R, López J. Bioremediation for a soil contaminated with hydrocarbons. Journal of Petroleum & Environmental Biotechnology. 2015;**6**:208

[66] US Environmental Protection Agency (EPA). How to evaluate alternative cleanup technologies for underground storage tank sites. In: A Guide for Corrective Action Plan Reviewers. Cincinnati, OH: Office of Research and Development. EPA 510-B-17-003; 1994

[67] Garima T, Singh SP. Application of bioremediation on solid waste management: A review. Journal of Bioremediation and Biodegradation. 2014;**5**(6):1-8

[68] Thomé A, Reginatto C, Cecchin I, Colla LM. Bioventing in a residual clayey soil contaminated with a blend of biodiesel and diesel oil. Journal of Environmental Engineering. 2015;**140**(11):1-6

[69] Sharma J. Advantages and limitations of in situ methods of bioremediation. Recent Advances in Biology and Medicine. 2019;**5**. Article ID: 955923

[70] Smitha MS, Singh S, Singh R. Microbial biotransformation: A process for chemical alterations. Journal of Bacteriology & Mycology: Open Access. 2017;**4**(2):47-51

[71] Jiang Y, Brassington KJ, Prpich G, Paton GI, Semple KT, Pollard SJT, et al. Insights into the biodegradation of weathered hydrocarbons in contaminated soils by bioaugmentation and nutrient stimulation. Chemosphere. 2016;**161**:300-307

[72] Størdal IF, Olsen AJ, Jenssen BM, Netzer R, Altin D, Brakstad OB. Biotransformation of petroleum hydrocarbons and microbial communities in seawater with oil dispersions and copepod feces. Marine Pollution Bulletin. 2015;**101**:686-693

[73] Karthikeyan R, Bhandari A. Anaerobic biotransformation of aromatic

and polycyclic aromatic hydrocarbons in soil microcosms: A review. Journal of Hazardous Substance Research. 2001;**3**:3-19

[74] Al-Bashir B, Cseh T, Leduc R, Samson R. Effect of soil/contaminant interactions on the bioremediation of naphthalene in flooded soil under denitrifying conditions. Applied Microbiology and Biotechnology. 1990;**34**(3):414-419

[75] Coste A, Micle V, Băbuţ CS. Study on the in-situ bioremediation techniques applicable to soil contaminated with petroleum products. ProEnviron. 2013;**6**:463-468

[76] Philp JC, Atlas RM. Bioremediation of contaminated soils and aquifers. In: Atlas RM, Philp JC, editors. Bioremediation: Applied Microbial Solutions for Real-World Environmental Cleanup. Washington: American Society for Microbiology (ASM) Press; 2005. pp. 139-236

[77] Daghio M, Aulenta F, Vaiopoulou E, Franzetti A, Arenda JBA, Sherry A, et al. Electrobioremediation of oil spills. Water Research. 2017;**114**:351-370

[78] Kao CM, Chen CY, Chen SC, Chien HY, Chen YL. Application of in situ biosparging to remediate a petroleum hydrocarbon spill spite: Field and microbial evaluation. Chemosphere. 2007;**70**:1492-1499

[79] Azubuike CC, Chikere CB, Okpokwasili GC. Bioremediation techniques—Classification based on site of application: Principles, advantages, limitations and prospects. World Journal of Microbiology & Biotechnology. 2016;**32**(11):180

[80] Megharaj M, Naidu R. Soil and brownfield bioremediation. Microbial Biotechnology. 2017;**10**(5):1244-1249

[81] Zappi ME, Bajpai R, Hernandez R, Taconi K, Gang D. Reclamation of smaller volumes of petroleum hydrocarbon contaminated soil using an innovative reactor system: A case study evaluation of the design. Agricultural Sciences. 2017;**8**:600-615

[82] McCauley P, Glaser J. Slurry Bioreactors for Treatment of Contaminated Soils, Sludges, and Sediments. Seminar on Bioremediation and Hazardous Waste Sites: Practical Approaches to Implementation, EPA/625/K-96/001. Vol. 12. Washington, DC: US EPA, Office of Research and Development; 1996. pp. 1-9

[83] Pino-Herrera DO, Pechaud Y, Huguenot D, Esposito G, van Hullebusch ED, Oturan MA. Removal mechanisms in aerobic slurry bioreactors for remediation of soils and sediments polluted with hydrophobic organic compounds: An overview. Journal of Hazardous Materials. 2017;**339**:427-449

[84] Robles-González IV, Fava F, Poggi-Varaldo HM. A review of slurry bioreactors for bioremediation of soils and sediments. Microbial Cell Factories. 2008;**7**:5

[85] Zhang C, Hughes JB, Nishino SF, Spain JC. Slurry-phase biological treatment of 2,4-dinitrotoluene and 2,4-dinitrotoluene: Role of bioaugmentation and effects of high dinitrotoluene concentrations. Environmental Science & Technology. 2000;**34**(13):2810-2816

[86] NSW EPA. Best Practice Note: Landfarming. State of New State Wales and Environmental Protection Authority; 2014

[87] Tuhuloula A, Altway A, Juliastuti SR, Suprapto S. Biodegradation of soils contaminated with naphthalene

in petroleum hydrocarbons using bioslurry reactors. IOP Conference Series: Earth and Environmental Science. 2018;**175**:012014

[88] Brown DM, Okoro S, Van Gils J, Van Spanning R, Bonte M, Hutchings T, et al. Comparison of landfarming amendments to improve bioremediation of petroleum hydrocarbons in Niger Delta soils. Science of the Total Environment. 2017;**596-597**:284-292

[89] Wang S, Wang X, Zhang C, Li F, Guo G. Bioremediation of oil sludge contaminated soil by landfarming with added cotton stalks. International Biodeterioration and Biodegradation. 2016;**106**:150-156

[90] Guarino C, Spada V, Sciarrillo R. Assessment of three approaches of bioremediation (natural attenuation, landfarming and bioaugmentation–assisted landfarming) for a petroleum hydrocarbon contaminated soil. Chemosphere. 2017;**170**:10-16

[91] Mphekgo P, Maila, Cloete TE. Bioremediation of petroleum hydrocarbons through landfarming: Are simplicity and cost-effectiveness of the only advantages? Reviews in Environmental Science and Biotechnology. 2004;**3**:349-360

[92] Benyahia F, Embaby AS. Bioremediation of crude oil-contaminated desert soil: Effect of biostimulation, bioaugmentation and bioavailability in biopile treatment systems. International Journal of Environmental Research and Public Health. 2016;**13**(2):219

[93] Wu G, Coulon F. Protocol for biopile construction treating contaminated soils with petroleum hydrocarbons. In: McGenity T, Timmis K, Nogales B, editors. Hydrocarbon and Lipid Microbiology Protocols. Springer, Berlin, Heidelberg: Springer Protocols Handbooks; 2015

[94] Raju MN, Scalvenzi L. Petroleum degradation: Promising biotechnological tool for bioremediation. In: Recent Insights in Petroleum Science and Engineering. London, UK: IntechOpen; 2018

[95] Whelan MJ, Coulon F, Hince G, Rayner J, McWatters R, Spedding T, et al. Fate and transport of petroleum hydrocarbons in engineered biopiles in polar regions. Chemosphere. 2015;**131**:232-240

[96] Kim J, Lee AH, Chang W. Enhanced bioremediation of nutrient-amended petroleum hydrocarbon contaminated soil over a cold climate winter: The rate and extent of hydrocarbon biodegradation and microbial response in a pilot-scale biopile subjected to natural seasonal freeze-thaw temperatures. Science of the Total Environment. 2018;**612**(15):903-913

[97] Dias RL, Ruberto L, Calabró A, Balbo AL, Del Panno MT, Mac Cormack WP. Hydrocarbon removal and bacterial community structure in on-site bio stimulated biopile systems designed for bioremediation of diesel-contaminated Antarctic soil. Polar Biology. 2015;**38**:677-687

[98] Gomez F, Sartaj M. Optimisation of field scale biopiles for bioremediation of petroleum hydrocarbon contaminated soil at low temperature conditions by response surface methodology (RSM). International Biodeterioration and Biodegradation. 2014;**89**:103-109

[99] Ren X, Zeng G, Tang L, Wang J, Wang J, Deng Y, et al. The potential impact on the biodegradation of organic pollutants from composting

technology for soil remediation. Waste Management. 2018;**72**:138-149

[100] Saum L, Jiménez MB, Crowley D. Influence of biochar and compost on phytoremediation of oil-contaminated soil. International Journal of Phytoremediation. 2018;**20**(1):54-60

[101] Cai D, Yang X, Wang S, Chao Y, Morel JL, Qiu R. Effects of dissolved organic matter derived from forest leaf litter on biodegradation of phenanthrene in aqueous phase. Journal of Hazardous Materials. 2017;**324**:516-525

[102] Atagana HI. Compost bioremediation of hydrocarbon-contaminated soil inoculated with organic manure. African Journal of Biotechnology. 2008;**7**(10):1516-1525

[103] Prakash V, Saxena S, Sharma A, Singh S, Singh SK. Treatment of oil sludge contamination by composting. Journal of Bioremediation & Biodegradation. 2015;**6**(3):284

[104] Coulon F, Wu G. Determination of petroleum hydrocarbons from soils and sediments using ultrasonic extraction. In: Hydrocarbon and Lipid Microbiology Protocol. Springer Protocols Handbooks. Berlin, Heidelberg: Springer; 2014

[105] Al-Daher R, Al-Awadhi N. Bioremediation of damaged desert environment using the windrow soil profile system in Kuwait. Environment International. 1998; **24**(1/2):175-180

[106] Shi Z, Liu J, Tang Z, Zhao Y, Wang C. Vermiremediation of organically contaminated soils: Concepts, current status, and future perspectives. Applied Soil Ecology. 2020;**147**:103377

[107] Njoku KL, Yussuf A, Akinola MO, Adesuyi AA, Jolaoso AO, Adedokun AH. Mycoremediation of petroleum hydrocarbon polluted soil by *Pleurotus pulmonarius*. Ethiopian Journal of Environmental Studies and Management. 2016;**9**(Suppl. 1):865-875

[108] Chachina SB, Voronkova NA, Baklanova ON. Biological remediation of the petroleum and diesel contaminated soil with earthworms *Eisenia fetida*. Process Engineering. 2016;**152**:122-133

[109] Dabke SV. Vermiremediation of heavy metal-contaminated soil. Blacksmith Institute Journal on Health and Pollution. 2013;**3**(4):4-10. DOI: 10.5696/2156-9614-3.4.4

[110] Ekperusi OA, Aigbodion FI. Bioremediation of petroleum hydrocarbons from crude oil contaminated soil with the earthworm: *Hyperiodrilus africanus*. 3 Biotech. 2015;**5**:957

[111] Schaefer M, Juliane F. The influence of earthworms and organic additives on the biodegradation of oil contaminated soil. Applied Soil Ecology. 2007;**36**(1):53-62

[112] Sanchez-Hernandez JC. Vermiremediation of pharmaceutical-contaminated soils and organic amendments. In: Pérez Solsona S, Montemurro N, Chiron S, Barceló D, editors. Interaction and Fate of Pharmaceuticals in Soil-Crop Systems. Handbook Environ. Chem. Vol. 103. Cham: Springer; 2020

[113] Azizi AB, Liew KY, Noor ZM, Abdullah N. Vermiremediation and mycoremediation of polycyclic aromatic hydrocarbons in soil and sewage sludge mixture: A comparative study. International Journal of Environmental Science and Development. 2013; **4**(5):565-568

[114] Sinha RK, Bharambe G, Ryan D. Converting wasteland into wonderland by earthworms – A low-cost nature's technology for soil remediation: A case study of vermiremediation of PAHs contaminated soil. The Environmentalist. 2008;**28**:466-475

[115] Kumar A, Tripti, Prasad MNV, Maiti SK, Favas PJC. Mycoremediation of mine site rehabilitation. In: Bio-Geotechnologies for Mine Site Rehabilitation. Elsevier BV. 2018. pp. 233-260

[116] Ali A, Guo D, Mahar A, Wang P, Shen F, Li R, et al. Mycoremediation of potentially toxic trace elements—A biological tool for soil cleanup: A review. Pedosphere. 2017;**27**(2):205-222

[117] Anderson C, Juday G. Mycoremediation of petroleum: A literature review. Journal of Environmental Science & Engineering. 2016;**A5**:397-405

[118] Barh A, Kumari B, Sharma S, Annepu SK, Kumar A, Kamal S, et al. Chapter 1—Muchroom mycoremediation: Kinetics and mechanism. Smart Bioremediation Technologies: Microbial Enzymes. 2019:1-22

[119] Jang KY, Cho SM, Seok SJ, Kong WS, Kim GH, Sung JM. Screening of biodegradable function of indigenous ligno-degrading mushroom using dyes. Mycobiology. 2009;**37**(1):53

[120] Bhatnagar S, Kumari R. Bioremediation: A sustainable tool for environmental management—A review. Annual Research & Review in Biology. 2013;**3**(4):974-993

[121] Yamada-Onodera K, Mukumoto H, Katsuyaya Y. Degradation of polyethylene by a fungus, *Penicillium simplicissimum* YK. Polymer Degradation and Stability. 2001;**72**:323-327

[122] Aranda E, Scervino JM, Godoy P, Reina R, Ocampo JA, Wittich R-M, et al. Role of arbuscular mycorrhizal fungus *Rhizophagus custos* in the dissipation of PAHs under root-organ culture conditions. Environmental Pollution. 2013;**181**:182-189

[123] Ulfig K, Płaza G, Worsztynowicz A, Mańko T, Tien AJ, Brigmon RL. Keratinolytic fungi as indicators of hydrocarbon contamination and bioremediation progress in a petroleum refinery. Polish Journal of Environmental Studies. 2003;**12**:245-250

[124] Phang SM, Chu WL, Rabiei R. Phycoremediation. In: Sahoo D, Seckbach J, editors. The Algae World. Cellular Origin, Life in Extreme Habitats and Astrobiology. Vol. 26. Dordrecht: Springer; 2015

[125] Gani P, Sunar NM, Matias-Peralta P, Latiff AA, Parjo UK, Ab Razak AR. Phycoremediation of wastewaters and potential hydrocarbon from microalgae: A review. Advances in Environmental Biology. 2015;**9**(20):1-8

[126] Kushwaha D, Saha S, Dutta S. Enhanced biomass recovery during phycoremediation of Cr(VI) using cyanobacteria and prospect of biofuel production. Industrial and Engineering Chemistry Research. 2014;**53**(51):19754-19764

[127] Abd-Razak SB, Sharip Z. The potential of phycoremediation in controlling eutrophication in tropical lake and reservoir: A review. Desalination and Water Treatment. 2020;**180**:164-175

[128] Subashchandrabose SR, Ramakrishnan B, Megharaj M, Venkateswarlu K, Naidu RR. Mixotrophic cyanobacteria and microalgae as distinctive biological agents for

organic pollutant degradation. Environment International. 2013;**51**:59-72

[129] Fathi AA, Azooz MM, Al-Fredan MA. Phycoremediation and the potential of sustainable algal biofuel production using wastewater. American Journal of Applied Sciences. 2013;**10**(2):189-194

[130] Dwivedi S. Bioremediation of heavy metals by algae: Current and future perspective. Journal of Advanced Laboratory Research in Biology. 2012;**3**(3):195-199

[131] He PJ, Mao B, Shen CM, Shao LM, Lee DJ, Chang JS. Cultivation of *Chlorella vulgaris* on wastewater containing high levels of ammonia for biodiesel production. Bioresource Technology. 2013;**129**:177-181

[132] El-Sheekh MM, Hamouda RA, Nizam AA. Biodegradation of crude oil by *Scenedesmus obliquus* and *Chlorella vulgaris* growing under heterotrophic conditions. International Biodeterioration and Biodegradation. 2013;**82**:67-72

[133] Kalhor AX, Movafeghi A, Mohammadi-Nassab AD, Abedi E, Bahrami A. Potential of the alga *Chlorella vulgaris* for biodegradation of crude oil hydrocarbons. Marine Pollution Bulletin. 2017;**123**:286-290

[134] Ossai IC, Hassan A, Hamid FS. Remediation of soil and water contaminated with petroleum hydrocarbon: A review. Environmental Technology and Innovation. 2020;**17**:100526

[135] Shrirangasami SR, Rakesh SS, Murugaragavan R, Ramesh PT, Varadharaj S, Elangovan R, et al. Phytoremediation of contaminated soils—A review. International Journal of Current Microbiology and Applied Sciences. 2020;**9**(11): 3269-3283

[136] Favas P, Pratas J, Varun M, D'Souza R, Paul MS. Phytoremediation of soil contaminated with metals and metalloids at mining areas: Potential of native flora. In: Hernandez-Soriano MC, editor. Environmental Risk Assessment of Soil Contamination. London, UK: IntechOpen; 2014

[137] Cristaldi A, Conti GO, Jho EH, Zuccarello P, Grasso A, Copat C, et al. Phytoremediation of contaminated soils by heavy metals and PAHs. A brief review. Environmental Technology & Innovation. 2017; **8**:309-326

[138] Hussain I, Puschenreiter M, Gerhard S, Schöftner P, Yousaf S, Wang A, et al. Rhizoremediation of petroleum hydrocarbon contaminated soil: Improvement opportunities and field applications. Environmental and Experimental Botany. 2018;**147**:202-219

[139] Gouda AH, El-Gendy AS, Abd El-Razek TM, El-Kassas HI. Evaluation of phytoremediation and bioremediation for sandy soil contaminated with petroleum hydrocarbons. International Journal of Environmental Science and Development. 2016;**7**(7):490-493

[140] Kösesakal T, Ünal M, Kulen O, Memon A, Yüksel B. Phytoremediation of petroleum hydrocarbons by using a freshwater fern species *Azolla filiculoides* lam. International Journal of Phytoremediation. 2015; **18**(5):467-476

[141] Muthusaravanan S, Sivarajasekar N, Vivek JS, Paramasivan T, Naushad M,

Prakashmaran J, et al. Phytoremediation of heavy metals: Mechanisms, methods and enhancements. Environmental Chemistry Letters. 2018;**16**:1339-1359

[142] Kvesitadze G, Khatisashvili G, Sadunishvili T, Ramsden JJ. Biochemical mechanisms of detoxification in higher plants. In: Basis Phytoremed. Vol. 4. Berlin Heidelberg: Springer, Verlag; 2006. pp. 185-194

[143] Cook RL, Hesterberg D. Comparison of trees and grasses for rhizoremediation of petroleum hydrocarbons. Inter. J. Phytoremed. 2013;**15**:844-860

[144] Dadrasnia A, Agamuthu P. Biostimulation and monitoring of diesel fuel polluted soil amended with biowaste. Petroleum Science and Technology. 2014;**32**:2822-2828

[145] Cartmill AD, Cartmill DL, Alarcón A. Controlled release fertiliser increased phytoremediation of petroleum-contaminated sandy soil. International Journal of Phytoremediation. 2014;**16**:285-301

[146] Agamuthu P, Abioye PO, AbdulAziz AR. Phytoremediation of soil contaminated with used lubricant oil using *Jatropha curcas*. Journal of Hazardous Materials. 2010;**179**(1-3):891-894

[147] Srikanth S, Kumar M, Puri SK. Bio-electrochemical system (BES) as an innovative approach for sustainable waste management in petroleum industry. Bioresource Technology. 2018;**265**:45-51

[148] Wang H, Luo H, Fallgren PH, Jin S, Ren ZJ. Bioelectrochemical system platform for sustainable environmental remediation and energy generation. Biotechnology Advances. 2015;**33**(3-4): 317-334

[149] Saxena G, Thakur IS, Kuma V, Shah MP. Electrobioremediation of contaminants: Concepts, mechanics, applications and challenges. In: Shah M, Banerjee A, editors. Combined Application of Physico-Chemical and Microbiological Processes for Industrial Effluent Treatment Plant. Singapore: Springer; 2020

[150] Lai A, Verdini R, Aulenta F, Majone M. Influence of nitrate and sulfate reduction in the bioelectrochemically assisted dechlorination of cis-DCE. Chemosphere. 2015;**125**:147-154

[151] Zhang T, Tremblay PL, Chaurasia AK, Smith JA, Bain TS, Lovley DR. Anaerobic benzene oxidation via phenol in *Geobacter metallireducens*. Applied and Environmental Microbiology. 2013;**79**:7800-7806

[152] Srikanth S, Kumar M, Singh D, Singh MP, Das BP. Electro-biocatalytic treatment of petroleum refinery wastewater using microbial fuel cell (MFC) in continuous mode operation. Bioresource Technology. 2016;**221**:70-77

[153] Palma E, Daghio M, Franzetti A, Papini MP, Aulenta F. The bioelectric well: A novel approach for *in situ* treatment of hydrocarbon-contaminated groundwater. Microbial Biotechnology. 2017;**6**:425-434

[154] Sinha S, Mehrotra T, Srivastava A, Srivastava A, Singh R. Nanobioremediation technologies for potential application in environmental cleanup. Environmental Biotechnology. 2020;**2**

[155] Galdames A, Mendoza A, Orueta M, deSoto-Gracia IS, Sánchez M, Virto I, et al. Development of new remediation technologies for contaminated soils based on the application of zero-valent iron nanoparticles and bioremediation

with compost. Resource-Efficient Technologies. 2017;**3**(3):166-176

[156] Yadav KK, Singh JK, Gupta N, Kumar V. A review of nanobioremediation technologies for environmental cleanups: A novel biological approach. Journal of Materials and Environmental Science. 2017;**8**(2):740-757

[157] Koul B, Taak P. Nanobioremediation. In: Biotechnological Strategies for Effective Remediation of Polluted Soils. Singapore: Springer; 2018. pp. 197-220

[158] Kumari B, Singh DP. A review on multifaceted application of nanoparticles in the field of bioremediation of petroleum hydrocarbons. Ecological Engineering. 2016;**97**:98-105

[159] Thomé A, Reddy KK, Reginatto C, Cecchin I. Review of nanotechnology for soil and groundwater remediation: Brazilian perspectives. Water, Air, & Soil Pollution. 2015;**226**:1-20

[160] Cecchin I, Reddy KI, Thome A, Tessaro EF, Schnaid F. Nano bioremediation: Integration of nanoparticles and bioremediation for sustainable remediation of chlorinated organic contaminants in soils. International Biodeterioration and Biodegradation. 2017;**119**:419-428

[161] Nnaji JC. Nanomaterials for remediation of petroleum contaminated soil and water. Umudike Journal of Engineering and Technology. 2017;**3**(2):23-29

[162] Otto M, Floyd M, Bajpai S. Nanotechnology for site remediationNanotechnology for site remediation. Remediation. 2008;**19**(1):99-108

[163] Karn B, Kuiken T, Otto M. Nanotechnology and in situ remediation: A review of benefits and potential risks. Environmental Health Perspectives. 2009;**117**(2):1823-1821

[164] Vázquez-Núñez E, Molina-Guerrero CE, Peña-Castro JM, Fernández-Luqueño F, de la Rosa-Álvarez M-G. Use of nanotechnology for the bioremediation of contaminants: A review. PRO. 2020;**8**:826

[165] Zhang WX. Nanoscale iron particles for environmental remediation: An overview. Journal of Nanoparticle Research. 2003;**5**:323-332

[166] Theron J, Walker JA, Cloete TE. Nanotechnology and water treatment: Applications and emerging opportunities. Critical Reviews in Microbiology. 2008;**34**:43-69

[167] NanoRem 1. Nanotechnology for Contaminated Land Remediation – Possibilities and Future Trends Resulting from the NanoRem Project. CL: AIRE NanoRem Bulletin; 2016

[168] Reddy KR, Darko-Kagya K, Cameselle C. Electrokinetics-enhanced transport of lactate modified nanoscale iron particles for degradation of dinitrotoluene in clayey soils. Separation and Purification Technology. 2011;**79**:230-237

[169] Ulfig K, Przystaś W, Płaza G, Miksch K. Biodegradation of Petroleum Hydrocarbons by Keratinolytic Fungi. In: Twardowska I, Allen HE, Häggblom MM, Stefaniak S, editors. Soil and Water Pollution Monitoring, Protection and Remediation, NATO Science Series. Vol. 69. Dordrecht: Springer; 2006

[170] Ingole NW, Vinchurkar SS, Dharpal SV. Adsorption of oil from wastewater by using human hairs. Journal of Environmental Science, Computer Science and

Engineering & Technology. 2013;**3**(1):207-217

[171] Pagnucco R, Philips ML. Comparative effectiveness of natural by-products and synthetic sorbents in oil spill booms. Journal of Environmental Management. 2018;**225**:10-16

[172] Ukotije-Ikwut PR, Idogun AK, Iriakuma CT, Aseminaso A, Obomanu T. A novel method for adsorption using human hairs as a natural oil spill sorbent. International Journal of Scientific & Engineering Research. 2016;**7**(8):1755-1765

[173] Cervantes-Gonzalez E, Rojas-Avelizapa LI, Cruz-Carmarillo R, Rojas-Avelizapa NG, Garcia-Mena J. Feather waste as petroleum sorbent: A study of its structural biodegradation. Proceedings of the Annual International Conference on Soils, Sediments, Water and Energy. 2010;**13**(7):50-58

[174] Koshlaf E, Ball AS. Soil bioremediation approaches for petroleum hydrocarbon polluted environments. AIMS Microbiology. 2017;**3**(1):25-49

[175] Salleh AB, Ghazali FM, Abd Rahman RNZ, Basari M. Bioremediation of petroleum hydrocarbon pollution. Indian Journal of Biotechnology. 2002;**2**:411-425

Section 3

Recovery of Value Added Compounds

Chapter 5

Hazardous Waste Granule Composting by Cycled Retort Using Microwave Radiated Asphalt/ Asphaltite Coal Slime Mixing

Yıldırım İsmail Tosun

Abstract

The hazardous sludge metal content of Mazıdağı metal leaching and electrowinning plants causes a great threat to ecology. The high-level metal and salt contaminants occurred in the copper leaching waste tailing ponds. The seepage liquors leak through the permeable bottom of ponds, such as acidic seepages. While urbanization needs freshwater; freshwater demand in the region increases because of global warming and drought. The estimated contamination values are avoided designing the controlling contamination level systems and meeting the disposal compost demand. In this approach, the demand for land covering and compost disposal has been designed as pellet or granule units determined independently of the specific needs of fertilizer products, agricultural remediation, and human needs. The amounts of sludge and wet hazardous toxic waste sludge's of Mazıdağı Phosphate Plants of Eti Bakır in Mardin change the ecosystem. The hazardous sludge of plant tailings is needed planned paste disposal or controlled regional dumping, pool effluents barrier on regarding seepage control demand of the freshwater lake of town region. The planned work is disposal tests for waste sludge composting as pasting. Additionally, this method protects ecology and improves waste sludge disposal by neutralizing it at a small scale. Even the toxicity will be easily monitored. Heavy metal contamination hazard maps will be prepared and an agricultural warning system will be established for agricultural irrigation.

Keywords: pasting heavy metals, microwave melted asphalt, shale, biochar, waste granule, hazardous industrial waste, sludge, road pavement

1. Introduction

Microwave radiation creates energy increase in the mineral crystals by friction so as the material body can partially permit radiation. Hence, thermal body heat increases with mechanical vibration stress. In that circumstance, the mineral crystals can generate thermal crack space regarding the permittivity and contents. In this study, the easy heating to 90°C, over the effect of microwave energy managed the melting waste sludge with asphalt bitumen contents in the retort reactor. By the melting character of asphalt in the furnace during that process, Şırnak asphaltite slime, a type of bitumen coal was investigated for compaction change and compressing period.

Aggregate is the main raw material of road superstructures as in many construction manufacturing. The properties of the aggregate, which forms a large part of the bituminous hot mix used in the superstructure, greatly affect the mixing performance. Therefore, it may be difficult to obtain aggregates that provide the required properties for BSKs and the aggregate costs are higher [1–4].

Hazardous radioactive or wet industrial sludge wastes processed for the recovery of metal with sulfuric acid, and methods for oxidized products with decay for acidic harmful substances. Thus, both soil wastes, complex carbon of sludge, and agricultural biomass wastes are evaluated. As well as other organic materials such as iron slime material and cellulosic material, they can be evaluated for high wetness. This waste was mixed by cleaning the acid mixture of the waste industry iron and chromium foundry sludge by pelletizing it with salt and mixing with the pelletized to cellulosic with exposed to decay time. This alkali salt is further salted then trapped. Disposal of sludge covers for scrapping has been examined on the impervious barrier at the bottom of large pools. As the coating of the bitumen mixture of the block pond size of these sludge wastes, the acidic solution with a solids content of 11.5% and hazardous radiation and the asphalt covering 10 mm hazardous wastewater pellets were degraded there in the dimensions of 10 mm. With this storage method, water resources are conserved and agricultural settings are also advantageous for improvement and environment. In our country, it is spread over a wide area outside the wide rocky soils in Siirt and Hakkari. The seepage of heavy metal ions or acidic chelate, mixing to streams should be neutralized by oxidizing reagents such as ozone or neutralizing alkaline washing so that it is reduced effluent contamination by high levels of heavy metals. The waste leaching liquors such as Hg, Pb, Cr, Cd, Cu, Zn, Fe, SO4 rates were contaminated freshwater sources near urbanization areas. The oxidation recycling of residual fly ash contaminated waters was a serious threat to the radioactive level. The hazardous chemical substance reacting that deteriorated environment was eliminated by pasting asphalt mixture and the following use as road pavement or filler source for waterproof barrier construction (**Figure 1**).

The Mazıdağı urbanization fieşd and freshwater dam lake, Mardin city is getting high ecological water safety threat on its population and density of immigrants by metal leaching phosphoric acid and sulfuric acid usage increase, its seepage, soil quality, and quantity of heavy metal in the soil. The main purpose for hazardous waste management in the Mazıdağ Phosphoric Acid and Electrowining Plant is easily controlled deposition of waste sludges without affecting water contamination. The water treatment plant in Mardin Eti Bakır Mazıdağı Plant will improve drinking water quality to be preserved. However, tailings containing hematite and iron sulfide with a high amount of sulfate and phosphate salts and tailing ponds water greatly threaten the ecology and near agricultural fields with a lack of irrigation.

1.1 Asphalt composting

The compost sludge volume was 10–20% in pasting, 90–95% of the volume was compost filler matters, 15% of the volume consists of fly ash fines [5–8]. The primary or secondary cause of compaction in paste and asphalt composts deteriorates the use of unsuitable coarse aggregates or the use of aggregates containing undesirable shaped substances [7–14].

Type of aggregate surface roughness, flatness, gradation, such as the characteristics of the bitumen binder thaw fatigue and tire performance has great importance on the

Figure 1.
The 3D view of Mazıdağı, Mardin. EtiBakır Mazıdağı phosphoric acid, sulfuric acid, metal leaching, and metal electrowinning plants.

asphalt composite track [14–21]. Aggregate-asphalt mixtures must meet certain requirements to be used as a pavement. These conditions are given in the Technical Specifications for coatings [21–24]. Natural stone crush aggregates are preferred in road superstructure construction in our country. But in many countries, natural aggregates as well as, artificial aggregates have been used in road construction. Aggregates are grouped as light, normal, and heavy aggregate considering the unit weight. These aggregates used in road construction in our country are mostly included in the normal and heavy aggregate class. In recent years, the use of artificially obtained light aggregates in road construction has found application in some countries.

The use of microwave melting in the production of asphalt composts reduces the use of cheap filler waste raw material resources, even the removal of sludge waste and controlled dumping requires substantial environmental benefits. In addition, the usage of fly ash in the asphalt compost production process allows the recovery of non-hazardous waste materials; thus, freshwater management, ecological, and economic benefits [25–34].

In this study, first all experiments were carried out to determine the physical properties of natural crushed stone aggregates and fly ash to be used in the mixture [34]. Sludge waste-melted asphalt briquettes were produced using Şırnak quarry limestone and asphaltite slime, aggregate using limestone, and char as compost mixing aggregate

The area was chosen as tailing pond area about 500 m away from the quarry and the sludged matter is wet at that 14 and 23.1% water of the dumping site near the Mazıdağı freshwater lake. Tailing ponds of metal leaching unit is near freshwater dam lake. On this lack of water for irrigation and drinking water demand of plant suffers from a relatively high potential of instability of tailing effluents seepages to groundwater levels by flood hazard and heavy rains. Geolayers for tailing ponds is providing control of the heavy metal-containing effluents seepages.

Şırnak asphaltite slime in the evaluation of local natural filler allows the production of bitumen binding strength of paste [14–16]. Weak strength of paste of asphalt compost reduces the contamination by increased fly ash filler use [17–19]. Plant for pasting of heavy metal sludge and composting by coal waste slime of Şırnak will improve drinking water quality to be preserved in Mardin.

2. Methods

The sludge matter hazards are seen on the ground come to seepage over the agricultural and freshwater sources in the area. In this case, it is clear that toxic parameters on the sludges will be compacted in the laboratories, avoiding the permanent dissolution values over 325 ppm to below 413 ppm–52 ppm. In this study, the toxic metal contaminants at lower levels than the threat level are about below 10 ppm. However, it was over 210 mg/l with some near groundwater wells and wet soils with leaks by uncontrolled seepages near tailings pond in Mazıdağ EtiBakır Plant area.

The well water heavy metal contamination levels were determined as given in **Table 1**.

Effluent, mg/l	**Tailing ponds sludge1 effluent**	**Tailing ponds sludge2 effluent**	**Tailing ponds sludge1**	**Tailing ponds sludge2**	**Well 1 irrigation soil water**	**Well 2 irrigation soil water**	**Dam irrigation soil water**
Cu	780	780	12.3	14.11	70	4.71	4.71
Pb	158	158	23.2	12.58	18	5.7	5.2
Fe	403.3	403.3	59	93.3	43.3	60.62	67.62
K + Na	852	852	8.7	8.52	552	≥70	≥50
Cd	276.2	276.2	14.1	14.72	47	16	15
Mn	123	123	24.2	43.3	13	≤25	≤25
Zn	372	372	15.7	7.2	57	≤15	≤15
PO4	1010	1010	2.8	2.10	220	≤5	≤5
SO_4	5700	5700	1.9	0.67	670	≤15	≤15
Solid, ppm							
Cu	124.1	124.1	52.3	48.71	40.71		
Pb	312.5	312.5	23.2	24.53	11.53		
Fe	444.3	444.3	5.9	7.62	5.62		
K + Na	747.5	747.5	81.7	81.46	88.6	≥70	≥50
Cd	246.7	246.7	10.1	9.56	19.56		
Mn	272	272	1.5	3.02	1.02	≤5	≤5
Zn	3330	3330	2.4	2.41	2.41	≤5	≤5
PO4	11,065	11,065	2.8	2.44	2.44		
SO_4	5798	5798	1.9	0.37	0.55		

Table 1.
The metal contents of the tailing pond sludges of the Mazıdağ EtiBakır plant and the potential soil well waters.

Component %	Şırnak asphaltite slime	Şırnak shale	Şırnak asphaltite char	Mazıdağı waste sludge	Şırnak asphaltite coal fly ash
SiO2	43.1	50.50	50.50	60.13	41.4
Al2O3	13.3	12.61	14.61	17.22	18.1
Fe2O3	9.5	14.30	24.30	4.59	4.2
CaO	7.4	12.30	2.30	2.48	18.4
MgO	3.7	2.3	1.28	2.17	4.0
K2O	2.5	2.0	2.51	3.51	2.1
Na2O	1.3	1.3	1.35	4.35	1.5
Ign.Loss.	38.9	6.6	12.21	4.12	1.6
SO3	0.2	1.2	0.12	0.52	0.2

Table 2.
Chemical composition of fillers of char composite present in the hazardous sludge pasting sample.

Char sludge and asphalt composite compaction as the mining wastes was also decreased the permeability of bottom layers in the dumping mining landfill field. The chemical composition of asphalt paste compost matters is given in **Table 2**.

2.1 Microwave asphalt -char -sludge compost melting

As a result of the Marshall calculations to determine the optimum amount of asphalt and type for high compaction is made with all three correlations, it is clear that if the lower limits of the aggregate types given in the specification are used in the sludge asphalt paste mixtures, a low binder will be required and the cost of the mixture will decrease. Bitumen properties of the sludge asphalt paste are given in **Table 3.** The physical properties of each produced sample were determined. The water sorption and porosity of the samples were determined by a series of soil standard experiments. In addition, compost quality comparison is made considering the use and supply of the waste materials used.

This study determined optimum volume reduction by compost compaction at finer particle size fraction rates and distribution factors for compaction to packed density.

Bitumen properties	
ASTM test	**Ortalama Değerler**
Penetration (25°C)	60–70
Flaming point	180°C
Ignition point	230°C
Melting point	45.5°C
Ductility(5 cm/min)	>100 cm
Specific gravity	1034

Table 3.
The bitumen matter properties in the asphalt pasting sludge studies.

Test, 0–4.75 mm	Şırnak asphaltite slime	Şırnak asphaltite char mm	Şırnak fly ash	Standartlar
Water sorption (%)	3.54	1.63	0.81	ASTM C 127
Los angeles (%)	27	26	23–38	ASTM C 131
Fineness (%)	15.5	3.7	4.5	ASTM C 117
Organic matter mad.	—	—	—	ASTM C 40
Thawing (%)	11	9	7.7	ASTM C 88
Abrasion (%)	—	≥%50	≥%50	
Density (gr/cm^3)	2.576	2.642	2.677	ASTM C 127
Rough weight (gr/cm^3)	1.61	1.40	1.41	ASTM C 29
Compact weight (gr/cm^3)	1.91	1.62	1.64	ASTM C 29

Table 4.
Granule properties, fineness of Şırnak limestone compost fillers present in the composite sample regarding ASTM standards.

	Şırnak asphaltite slime	Şırnak asphaltite char	Şırnak fly ash
Density (gr/cm^3)	1.855	1.655	1.855
Rough weight (gr/cm^3)	1.32	1.23	1.32
Compact weight (gr/cm^3)	1.84	1.84	1.94
Water sorption (%)	14.5	15.5	17.5
Moisture (%)	0.2	0.1	0.2

Table 5.
Şırnak Asphaltite slime, char granule properties, fineness of Şırnak char and fly ash fillers present in the composite sample.

Finally, a compaction volume rate of 32% could be managed by microwave melting and following pressing under 3 tones load for 50 mm diameter mold.

The physical and mechanical strength values investigated are given in **Table 4** with respect to standards. The oil addition as easy pressing through the mold in compaction is optimized at below 150-micron particle size fractions. The bitumen content was recovered back with a 95.7% recovery rate in compacting fine solids as given in **Table 5** following microwave melting.

2.2 Grain size analysis

Ground samples are sieved as mentioned in the standard method and sedimentation test. The particle gradation and size distribution were determined according to the chart in **Figure 2**. **Figure 3** shows the gradation of asphalt paste fillers (ASTM C136) [11].

The sludge wastes in the Mazıdağ tailing pond are given in **Table 1**.

2.3 Experimental pasting melted asphalt content calculation

The gradation factors and formula is the oldest McKessen - Fricstad formula for calculating asphalt content. The proposed binder addition amount is determined by the formula given below for asphalt bitumen bound aggregate composts [11, 12].

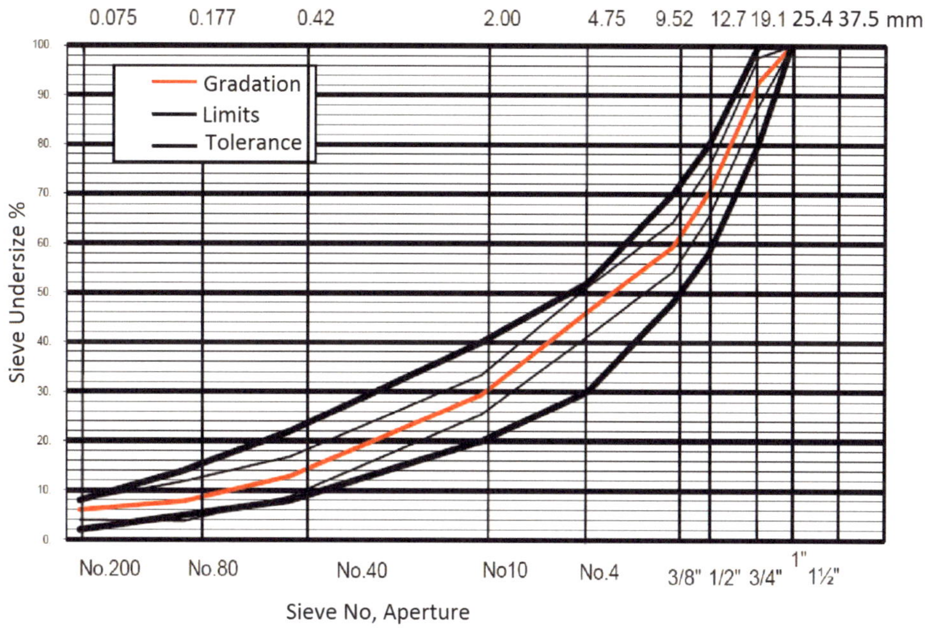

Figure 2.
All of the base and bottom base layers without binders are weighted by bituminous hot mixtures.

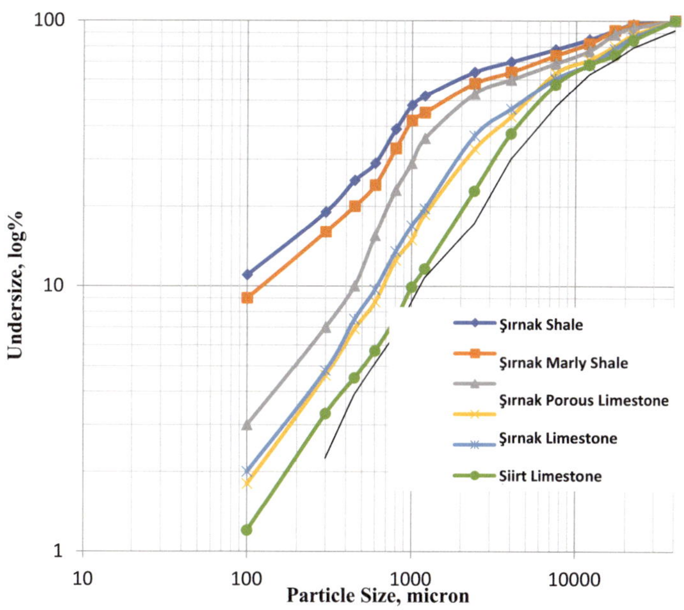

Figure 3.
Changes in Gaudin-Schumann particle size distribution of paste filler material.

$$p = 0,015\,P + 0,4\,S + 0,2\,F$$

P: Binder content in% by weight of aggregate a:% of aggregate remaining on sieve No 10.

S: 1% by weight of aggregate remaining between sieves No. 10 and 200.
F: % by weight of the aggregate passing through the sieve No.200.

The amphoric formula is commonly used in the gradation calculation of the compacting paving rock with asphalt. The bitumen content in asphaltic mixtures was determined by the following correlation.

$$SI = k\,\gamma\,\sqrt[n]{0,015\,C + 0,4\,S} + 0,3\,P$$

P: Asphalt content of % by weight of aggregate specific gravity of aggregate distribution factor α = 2.65/γ

$$\alpha = 0,015\,n + 0,4\,F$$

γ: Aggregate specific gravity.
k: Aggregate wealth module (3 to 3.5 coefficient).
F: Aggregate fine content %.
S: specific surface area of aggregate (m^2/kg).
Total Shear strength = 0.25G + 2.3P + 12SI + I35F.
G:% by weight of aggregates greater than 6.3 mm.
S:% by weight of the aggregates remaining between 6.3 mm–0.315 mm (# 50).

As it is seen in both experimental methods and empirical relations, an important factor affecting asphalt content is the aggregate gradation used to determine the asphalt content in asphaltic mixtures. The effect of aggregate gradation on the asphalt content was investigated in three parts. In the first part, binding contents of binder and wear layers aggregate gradations were found. In the second part, we investigated how aggregates passing through screens 1/2 ", 4, and 8 affect the contents of the binder.

The asphalt contents required for the lower and upper limits of the compost gradations are illustrated in **Figure 2** using the ASTM standards [14]. In **Figure 2**, the aggregate gradation for road pavement requiring at least asphalt content for the binder layer is the lower limit of type 3 providing high stability strength under layer. The upper limit of type 2 is within the upper limits providing high wear resistivity on the road pavement. For the ductility, it is type 1, which requires the medium binding content among the lower and upper limits of gradation.

2.4 Los Angeles values

The standard abrasion criteria of aggregates are determined by standard tumbling tests giving values of shattering by water and tumbling act [15]. The Şırnak asphaltite, shale, and char show low durability values in Los Angeles values among 40–55%. Regarding grain size of limestone aggregates, the compost properties are given in **Table 6**.

2.5 Microwave melted, compacted/briquetted Sludges

Stability is a significant parameter of compost impermeability for sludge paste avoiding contact with water sources in the landfill. The lowest asphalt binder at 3.5–4% weight rate in the compost increase the wear resistivity and stability high enough to resist compaction loads and contact to potential water seepages.

	Şırnak limestone			
Size, mm	**0–4.75**	**4.75–9.5 mm**	**9.5–25**	**Standards**
Water (%)	* (3.54)	1.63	0.81	ASTM C 127
Los Angeles (%)	44	55	73.804	ASTM C 131
Fine (%)	11.3	1.27	0.45	ASTM C 117
Organic binder	Berrak	Berrak	Berrak	ASTM C 40
Thaw. (%)	*	*	6.69	ASTM C 88
Abrasion. (%)	*	%50 'den Fazla	%50 'den Fazla	
γ Density (gr/cm^3)	2.576	2.642	2.677	ASTM C 127
Rough density γk gr/cm^3)	1.61	1.40	1.41	ASTM C 29
Compacted density (gr/cm^3)	1.91	1.62	1.64	ASTM C 29

Table 6.
The physical parameters of Şırnak limestone in paste compost.

However, very high stability means a very hard pasting mixture even a high amount of fine fillers for sludge pasting was required. On the other side ductility and durability decrease for the pastes by a high amount of thermal or stress cracking. The microwave heating provides the low asphalt content use in pasting and even easier compacting in this study. Although the asphalt leak of 0.1% occurred in the tests gasoil was used emulsification of asphalt due to reducing heterogeneous mixing and easy wetting. The high compaction of sludge and use of fly ash at 20–25% volume rate increased compaction ability and reduced the permeability. The permeability of sludge compost shows the resistance of the asphalt coating to the passage of air or water.

The permeability over 10^9 D is wished from waste pastes to avoid the hazardous effect of sludge contamination to the soil or water sources.

The impermeability is determined by the percentage of air voids in the mixture of low asphalt percentage and high voids in the design make the layer highly permeable.

By the way of microwave radiation at 10 minutes period increased compaction load to 1500 kg with 20 mm shear compaction distance in the barrel of 70 mm diameter of test mold. The followed process in sludge pasting is illustrated in **Figure 4**.

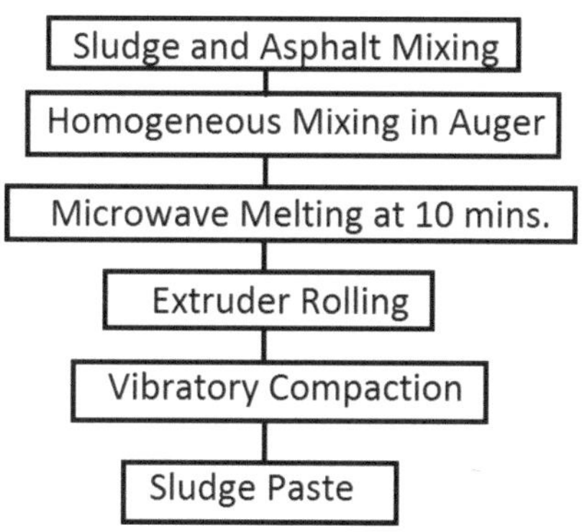

Figure 4.
Microwave experimentation flowsheet for sludge pasting with asphalt.

3. Results and discussion

3.1 Char compost compaction strength

The hazardous sludge threat is widened in similar areas. The urbanization districts which are at 1 km distance to Mazıdağı on the north of the Metal recovery plant and the tailing pond area of Mazıdağı Plant (**Figure 5**) were examined with permeability and hydrology of land in different locations at 100 m away. However, in this study char carbon reduces to contaminated soil and irrigation water sources by metal absorption ability. 20–10% volume rate of char in the compost will decrease dissolution amount to water at ten-fold times by homogeneous mixing under microwave melting radiation on surface interference. The compaction ability is even increased absorption amount by addition to char and coal slime to the pasting. The coverage binding bitumen and wetting effect of asphalt is improved by the addition of char carbon content of compost.

3.2 Fine fraction of the hazardous sludge compost

In the present study, the microwave melted mixture of a Sirnak asphaltite slime below 100 microns as a carbon sample with coal pyrite improved the microwave melting effect before compaction. This resulted in almost complete melting of the compost sludge and ash material as seen in **Figure 4**.

3.3 Compost granule gradation

The medium-coarse content at 4.5 mm–2 mm provided the well gradation and optimum pasting method at an amount 25% weight rate compost fillers as illustrated in **Figure 6**. The finer size below 0.2 mm increased 50% weight rate determines the lowest asphalt content of the paste gradation used in the sludge mixture depending on the surface area. In this method, we briefly follow the following mixture types as given in **Table 7**.

The binder percentages are about 4.5–5% weight rate corresponding to the optimum compaction value so as the percentage of the space specified in the compaction. The binder compaction space percentages are the percentage of asphalt volume collected. The arithmetic averages of the bulk content were found as the best compaction

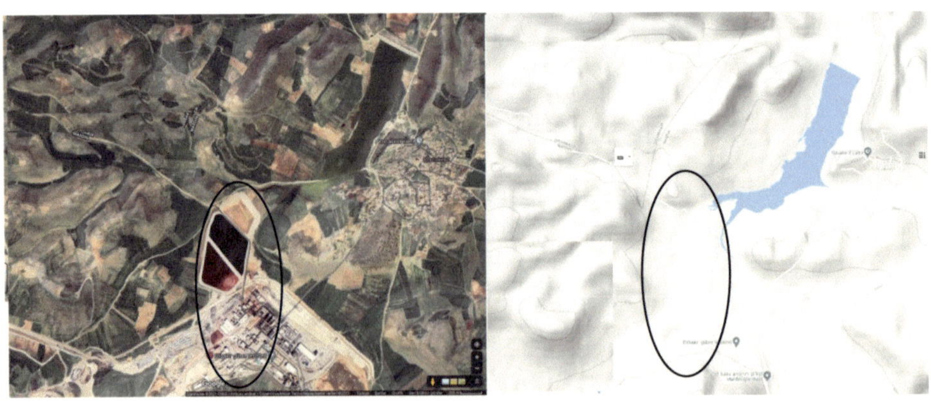

Figure 5.
Satellite view of Mazıdağı, Mardin. EtiBakır Mazıdağı metal leaching and electrowinning plants.

Figure 6.
Views of marl of Sirnak, limestone b, and c, d. after chemical interaction aggregate image.

Mixture component %	Sludge waste/ slag char	Sırnak limestone	Şırnak marl	Fly ash	Asphalt
A	65	3	2	25	10
A2	65	3	2	20	10
A3	55	5	5	20	15
C	65	4	1	20	10
C2	60	4	1	20	15
F	65	9	1	20	10
F2	60	9	1	20	15

Table 7.
Char, Şırnak asphaltite sludge compost paste filler matrix composition in Şırnak City.

asphalt content (the percentage of the aggregate weight in the mixture). This value is checked on the yield curve and the corresponding flow value is checked for compliance with the specification.

3.4 Compaction

The filler materials used in the pasting compost at 20% volume rate are viewed in **Figure 6**. Shale particles contained high porosity reaching 7.8% in the compost. The

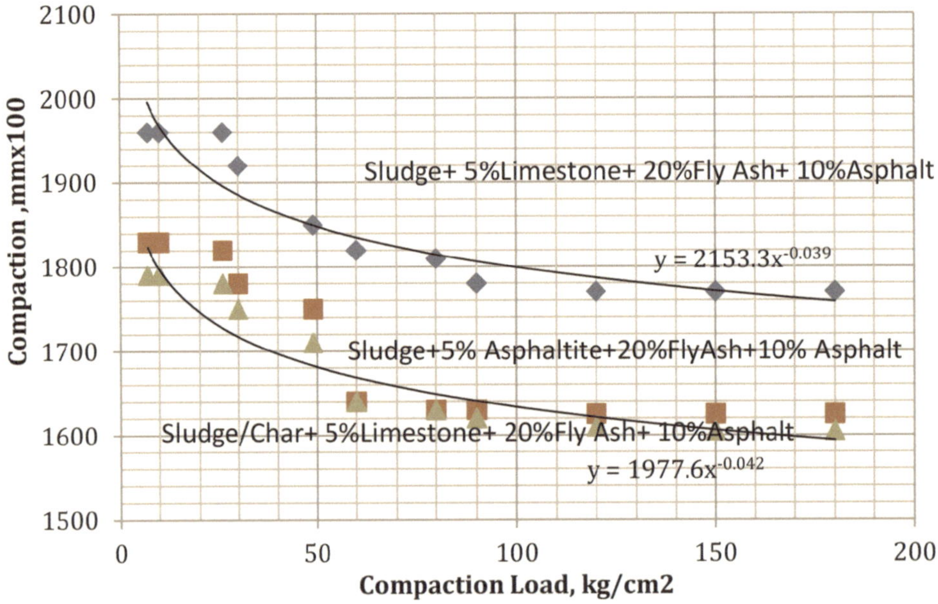

Figure 7.
Relative compaction changes of paste fill with Şırnak limestone, asphaltite slime char, and Şırnak shale after load indentation.

limestone's water absorption values were lower with asphaltite slime at about 2–3.5%. The filler mixing combinations are given in **Table** 7.

The effect of fly ash and Şırnak asphaltite and shale on compost density with asphalt mixing is illustrated in **Figure** 7.

In this study, the mechanical properties of local limestone and mine waste stone of Şırnak asphaltite slime and shale are used as filler for the purpose of asphalt sludge filler paste production. The waste slime asphaltite and Şırnak char are ground below 4 mm at the coarser part of 5% weight ratio. Şırnak shale and 15% Silopi fly ash are prepared as a binder cementing matter of sludge mixture of paste. The pressed blocks were manufactured in 50 × 50 × 10 mm like pile blocks, 10 mm thick bending strength, impact resistance, friction, abrasion loss, compaction distance were determined by the experiments. The ability of paste in the pile blocks to compact loss asphalt bitumen and water desorption are analyzed.

These experiments show that local rock cement mixtures and elaborate the application of cementing technique with silica fume used and it prevents the coating applied to moisturize the strengthening of historic buildings Sirnak region as a result and is determined to carry out the consolidation.

3.5 Shear load and compression strength analysis

It is determined that the 90°C temperature accelerates the melting of asphalt bitumen. These technological applications can be further developed with the evaluability of local natural stone powder in paste backfill. The strength of sludge paste pile blocks produced is reduced 7.2 to 6.4 MPa. Lower asphalt weight increased mechanical strength of pasted pile block with compaction been advantageous. Thus, the ideal compacting load of 100 kg tight aggregates with the block and pasted block resistance of the produced slag mixture may arise from 8.3 to 9.2 MPa.

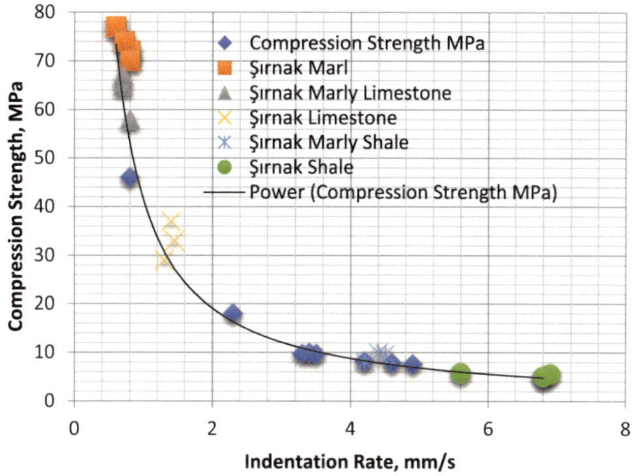

Figure 8.
Compression strength of paste fill material rocks.

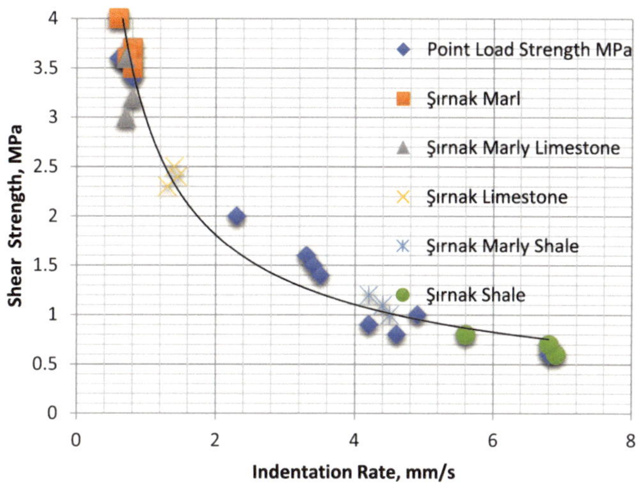

Figure 9.
Point load I strength of paste fill material rocks.

The results are shown in **Figures 8** and **9**. Indentation tests of the rock types were determined by a drill machine at 10 mm bit diameter due to the correlation of rock strengths underground mining cavity conditions.

3.6 Paste fill block production and compaction analysis

Asphalt and fly ash binders are raised to 10–45% weight ratio at different mixtures and 20% wet hazardous sludge and strength properties of prepared mixture paste with asphalt melted compost.

The water/binder (w/B) ratio is decided to be 20%/10–45% as a result of preliminary experiments. Pasting 5% melted asphalt is sufficient. The amount of each series fly ash composting fines are sorted asphalt and sludge in pile blocks and determined the compressive strength after drying blocks.

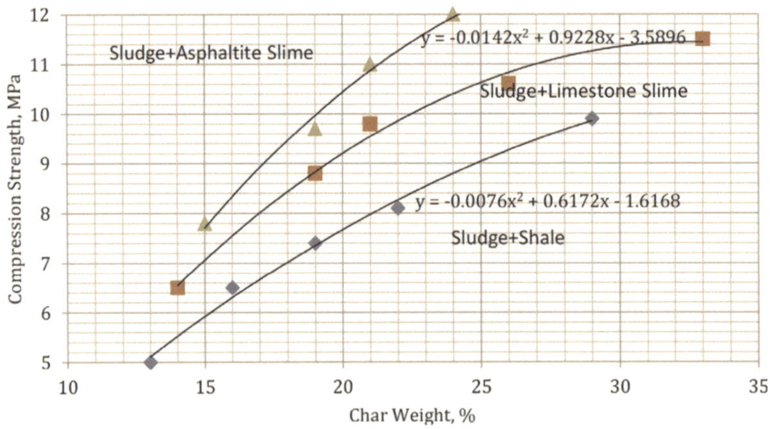

Figure 10.
Char addition weight of pile paste blocks with paste fill vs. UCS test results: 1. Shale 2. Sirnak limestone 3. Sirnak limestone.

4. Conclusions

Fly ash fine-texture and sludge mixing were homogeneous and the contamination act of sludge decreased at half amount rate by addition of 20% fly ash addition to paste mixture. The char carbon will also ten times decrease the metal contamination to water sources. The texture of low asphalt and emulsified hıgh bitumen content lowered the compaction resistance by increasing structure strength.

Especially the compact loading resistivity increased over 1500 kg for 70 mm barrel loading on which the resistance is kept as high as 20 mm displacement as high. Sand content of sludge was lower than the limestone so that this gradation provided high durability and less cracking during compaction loading final product briquettes and lower abrasion resistance.

Figure 10 shows the 1000 kg loading was suitable for long durability and optımum briquetting outputs. The stability was varied depending on the amount of fly ash and Şırnak asphaltite slime contained in the pasting relative to the pore contents in the micro-structural texture of asphalt compost. The pore density will decrease by low asphalt content and compaction resulting in the development of impermeable surface on the compost briquettes consequently.

The apparent uniaxial compressive strength test of Şırnak limestone showed sufficient substrate gradation and porous texture with high Los Angeles values of 67% at 4.5 mm granules. The gradation was suitable for high compaction ability and Şırnak limestone exhibited low porosity and water sorption 0.8%.

The higher compaction over 1000 kg provided sufficient permeability and compression strength values and higher shear resistance with lower cracking over time period of dynamical stress loads related to landfill application.

Symbols

τ kg / cm^2	Shear stress
σ kg / cm^2	Normal stress
Ip	Plasticity index

Ll	Liquid limit
Pl	Plastic limit
Wopt	Optimum water content
γ Natural g/cm^3	Natural unit volume weight
γsatg $/cm^3$	Saturated unit volume weight
γdry g/cm^3	Dry unit volume weight
γkmax g/cm^3	Maximum dry unit volume weight
γs g/cm^3	Grain unit volume weight
k	Permeability coefficient

Author details

Yıldırım İsmail Tosun
Engineering Faculty, Mining Engineering Department, Şirnak University, Şirnak, Turkey

*Address all correspondence to: yildirimismailtosun@gmail.com

References

[1] Anderson RM, Bahia HU. Evaluation and Selection of Aggregate Gradations for Asphalt Mixtures Using Superpave. In: Transportation Research Record 1583, TRB. Washington, D.C.: National Research Council; 1997. pp. 91-97

[2] Hinrichsen JA, Heggen J. Minimum Voids in Mineral Aggregate in Hot-Mix Asphalt Based on Gradation and Volumetric Properties. In: Transportation Research Record 1545, TRB. Washington, D.C.: National Research Council; 1996. pp. 75-79

[3] Kandhal PS, Foo KY, Mallick RB. A Critical Review of VMA Requirements in Superpave. NCAT Report 98-1. National Center for Asphalt Technology; 1998

[4] Kandhal PS, Chakraborty S. Evaluation of Voids in the Mineral Aggregate for HMA Paving Mixtures. NCAT Report 96-4. National Center for Asphalt Technology; 1996

[5] Hudson SB, Davis RL. Relationship of Aggregate Voidage to Gradation. Proc. Association of Asphalt Paving Technologists. 1965;**34**:574-593. https://trid.trb.org/Results?q=&serial=%22Assoc%20Asphalt%20Paving%20Technol%20Proc%22

[6] Richardson C. The Modern Asphalt Pavement. 1st ed. NewYork: John Wiley & Sons; 1905

[7] Anonymous. Mix Design Methods for Asphalt Concrete (MS-2). 2nd ed. USA: The Asphalt Institute; 1963

[8] Anonymous. Asphalt Mix Design Methods (MS-2). 7nd ed. USA: The Asphalt Institute; 2014

[9] Anonnymous. Highway Research Board, The Marshall Method for the Design and Control of Bituminous Paving Mixtures. 3rd rev ed. Miss. USA: Marshall Consulting and Testing Laboratory, Bulletine 105; 1955

[10] Arifin,S., M. Selintung, L Samang, 2015, Effect Of Effective Specific Gravity On Vma Of Asphalt Concrete (Case Study Of Lolioge Materials Of Palu City). ISSN: 2049-3444, IJET Publications UK. Vol 5, No 4ciation of Asphalt Paving Technologists, Vol. 24, April 1955

[11] McLeod NW. Relationship Between Density, Bitumen Content, and Voids Properties of Compacted Paving Mixtures. Proc., Highway Research Board, Vol. 35. In: Proceedings of the Thirty-Fifth Annual Meeting of the Highway Research Board, Washington, D.C. 1956

[12] Akbulut H, Çetin S, Gürer C. Andezit Agregasının Sıcak Karışım Asfalt Kaplamalarında Kullanılabilirliğinin Araştırılması, 5. Uluslararası İleri Teknolojiler Sempozyumu (IATS'09), Karabük. 2009

[13] ASTM C127 – 07. Standard Test Method for Density, Relative Density (Specific Gravity), and Absorption of Coarse Aggregate. Pennsylvania USA: Annual Book of ASTM Standards; 1992

[14] ASTM D5-97. Standard Test Method for Penetration of Bituminous Materials. Pennsylvania USA: Annual Book of ASTM Standards; 1997

[15] ASTM D70-03. Standard Test Method for Density of Semi-Solid Bituminous Materials (Pycnometer Method). USA: Annual Book of ASTM Standards Pennsylvania; 2003

[16] ASTM D36/36M-09. Standard Test Method for Softening Point of Bitumen

(Ring-and-Ball Apparatus). Pennsylvania USA: Annual Book of ASTM Standards; 2009

[17] ASTM D113-07. Standard Test Method for Ductility of Bituminous Materials. Pennsylvania USA: Annual Book of ASTM Standards; 2007

[18] ASTM D92-01. Standard Test Method for Flash and Fire Points by Cleveland Open Cup Tester. USA: Annual Book of ASTM Standards; 2001

[19] ASTM D 1559–89. Standard Test Method for Resistance to Plastic Flow of Bituminous Mixtures Using Marshall Apparatus. USA: Annual Book of ASTM Standards; 1989

[20] Yıldırım İT. Microwave extruded briquetting-pellets of coal slime and sewage sludge for humate, 4.Uluslararası Multidisipliner Kongresi. Girne, KKTC: Ekim; 2018. pp. 16-18

[21] Kreith F, Tchobanoglous G. Handbook of Solid Waste Management. NewYork: McGrawhill; 2002. DOI: 10.1007/978-981-15-7525-9

[22] Baskar C, Ramakrishna S, Baskar S, Sharma R, Chinnappan A, Sehrawat R. Handbook of Solid Waste Management. Singapore: Springer; 2022

[23] Sata V, Jaturapitakkul C, Kiattikomol K. Influence Of Pozzolan From Various By-Product Materials On Mechanical Properties Of High-Strength Concrete. Construction And Building Materials. 2007:1589-1598

[24] ASTM C 618-05 (2006) Specification for Coal Fly Ash and Raw or Calcined Natural Pozzolan for Use in Concrete. Annual Book of ASTM, Standard, Section 04 Construction, Volume 04.02 Concrete and Aggregate, ASTM International, West Conshohocken. www.astm.org.

[25] Bhattacharyya KG, Gupta SS. Adsorption of Fe(III), Co(II) and Ni(II) on ZrO-kaolinite and ZrO-montmorillonite surfaces in aqueous medium. Colloids Surf. A Physicochem. Eng. Asp. 2008;**317**:71-79. DOI: 10.1016/j.colsurfa. 2007. 09.037

[26] Abatal M, Quiroz VC, Olguín MT, Vázquez-Olmos AR, Vargas J, Anguebes-Franseschi F, et al. 2019, Sorption of Pb(II) from Aqueous Solutions by Acid-Modified Clinoptilolite-Rich Tuffs with Different Si/Al Ratios. Appl. Sci. 2019;**9**(12):2415. DOI: 10.3390/app9122415

[27] Xu X, Lin L, Papelis C, Xu P. Sorption of Arsenic from Desalination Concentrate onto Drinking Water Treatment Solids: Operating Conditions and Kinetics. Water 2018. 2019;**10**(2):96. DOI: 10.3390/w10020096

[28] Ofomaja AE, Naidoo EB, Modise SJ. Biosorption of copper (II) and lead (II) onto potassium hydroxide treated pine cone powder. J. Environ. Manage. 2010; **91**:1674-1685. DOI: 10.1016/j.jenvman.2010.03.005

[29] Bajpai SK, Jain A. Removal of copper (II) from aqueous solution using spent tea leaves (STL) as a potential sorbent. Water SA. 2010;**36**:221-228

[30] Chen H, Dai G, Zhao J, Zhong A, Wu J, Yan H. Removal of copper(II) ions by a biosorbent—Cinnamomum camphora leaves powder. J. Hazard. Mater. 2010;**177**:228-236. DOI: 10.1016/j.jhazmat.2009.12.022

[31] Bhattacharyya KG, Gupta SS. Adsorption of a few heavy metals on natural and modified kaolinite and montmorillonite: A review. Adv. Colloid

Interface. Sci. 2008;**140**:114-131. DOI: 10.1016/j.cis.2007.12.008

[32] Gupta SS, Bhattacharyya KG. Adsorption of Ni(II) on clays. J. Colloid. Interface. Sci. 2006;**295**:21-32. DOI: 10.1016/j.jcis.2005.07.073

[33] Barrera H, Ureña-Núñez F, Bilyeu B, Barrera-Díaz C. Removal of chromium and toxic ions present in mine drainage by Ectodermis of Opuntia. J. Hazard. Mater. 2006;**136**:846-853. DOI: 10.1016/j.jhazmat.2006.01.021

[34] Tosun Yİ. In: Eraslan M, Demirkaya H, Direkci B, Aslan F, Kilinç M, Kitapliği G, editors. Use of Coal Boiler Bottom Ash, Fly Ash and Geosynthethics for Bottom Seal Layer for Municipal Waste Landfill, Bölüm: Mühendislik. Ankara, Turkey. ISBN • 978-605-288-362-4, Çankaya/ ANKARA web: www.gecekitapligi.com: DİSİPLİNLERARASI AKADEMİK ÇALIŞMALAR- I; 2018. pp. 185-200

Chapter 6

Application of the Sewage Sludge in Agriculture: Soil Fertility, Technoeconomic, and Life-Cycle Assessment

Olga Muter, Laila Dubova, Oleg Kassien, Jana Cakane and Ina Alsina

Abstract

Disposal of sewage sludge, which is a by-product of wastewater treatment, has become one of the greatest challenges of the twenty-first century. Conversion of sewage sludge to a soil amendment can be performed by a broad spectrum of methods, which greatly differ by substrate/amendment composition, treatment time, and physicochemical conditions. The book chapter is focused on (i) environmental and legislative aspects of sewage sludge application in agriculture; (ii) risk factors related to the abundance of pathogens in sewage sludge and methods of SS hygienization; (iii) optimization of the use of SS-derived fertilizers. Application of sewage sludge in combination with mineral fertilizers positively influenced crop growth and soil microbiological activity. An environmental impact of sewage sludge related to its disposal to agricultural areas has been analyzed in terms of global warming, ecotoxicity, and other internationally recognized issues. Narrowly targeted measures may aggravate the situation. Some site-specific factors make sewage sludge unique, hence this specificity must be considered to predict the outcome of its treatment. Determination of these factors remains challenging. Therefore, the complexity of sewage sludge can be reduced by employing integrated biorefinery approaches that will result in circular bioeconomy and industrial ecology solutions.

Keywords: circular economy, fertilizer, life-cycle assessment, plant growth, sewage sludge hygienization

1. Introduction

Sewage sludge (SS) is formed as a by-product at a wastewater treatment plant (WWTP) and represents a heterogeneous mixture. This complex suspension consists of solid organic and inorganic substances and colloids, which have been separated from the wastewater during the treatment process [1]. The global production of SS is estimated at 45 million t of dry matter per year [2, 3]. During the last decade the SS production in EU countries increased by 1.5 million t of dry matter (DM), that is, from 11.5 million t in 2010 to 13 million t in 2020 [3], therefore, its management is a problem of great concern. The SS disposal reaches up to 60% of the total operating costs of WWTP, and, hence, makes this process problematic and expensive [4].

Sludge from WWTP is recovered by compost production, the application directly to agricultural and forest land, production of growing substrates, and energy recovery [5]. For practical and legal reasons, SS is increasingly reused rather than landfilled. This approach aims to minimize generated waste and promote the development of the bioeconomy that provides intelligent waste management, and, hence, is consistent with zero-waste strategy [3, 6]. Different countries have chosen different strategies for the use of urban SS. Analysis of the Eurostat data in the period from 2014 to 2018 showed that the use of SS in agriculture, in combination with compost, had been the main route for sludge disposal in the EU with 44.58%, followed by incineration (32.70%) and other methods of disposal (9.16%). Landfill disposal was at the level of 7.81%. Comparing the costs of different sludge disposal methods, the application on land and agriculture involves the lowest cost compared to composting, drying, incineration, and landfill.

At WWTPs, with more than 10,000 inhabitants, the sludge is divided into primary and secondary sludge. The primary sludge contains settling substances (from primary settling tanks), usually, it has a granular structure. Secondary sludge, also called excess sludge, consists of a mixture of microorganisms and settable substances from the biological stage of the WWTP. Primary sludge and secondary sludge are referred to as so-called raw sludge. The raw sludge is still microbially active, it can contain pathogenic microorganisms, with the total content of organic substances in the dry matter at about 70%. However, dewatered sludge (20–45% DM) is considered harmless and suitable for agriculture, because of high content of organic matter and biogenic elements (C, N, P), which increases soil fertility and is essential for plant growth and development as well as for soil microbiota [1]. Therefore, the use of SS on agricultural land is the best way to recycle the nutrients it contains, thus making the SS an important biological resource for sustainable agriculture [7–9]. On the other hand, the application rate is of great importance. Excessive concentrations of plant nutrients, mainly nitrogen and phosphorus, can also harm the environment, especially inland waters.

Another important issue is related to the abundance of hazardous and very persistent substances, such as heavy metals, polycyclic aromatic hydrocarbons, polychlorinated biphenyls, halogenated hydrocarbons, polychlorinated dibenzo-p-dioxins and dibenzofurans, pesticides, personal care products, hormonal substances, drugs and their metabolites, microplastics, and nanoparticles [8].

Therefore, the incorporation of sludge and its compost in the soil is regulated by various legislative acts [9]. The annual emission limit values for dry matter, heavy metals, total nitrogen, and total phosphorus are the maximum mass of these substances that can be applied per hectare of sludge or compost on average per year. Emission limit values for sludge dry matter vary considerably between the EU Member States, ranging from 1 to 10 t ha^{-1} per year. According to Mercl et al. [10], a high rate of SS composts applied once (60 t ha^{-1} compost in seedbed) is not recommendable since high nitrate concentration is not taken up by maize and increases the leaching risk. Furthermore, SS commonly contains high amounts of human pathogenic bacteria excreted in feces and urine, so the SS should be appropriately hygienized before application in agriculture.

The aim of this chapter was to summarize the main aspects of SS treatment for its application in agriculture, with emphasis on process efficiency, safety, and feasibility.

The dual role of SS as a fertilizer and amendment in the soil is widely described in the literature, referring to the supply of nutrients to plants and improving the soil's physical conditions, respectively. Our own results on SS treatment have been incorporated into the review of recent scientific literature and legislative documents.

2. Characteristics of sewage sludge as a potential soil amendment/ fertilizer

Sludge is rich in organic matter, nitrogen, phosphorus, and other macro and microelements, which makes it a useful raw material to be used in agriculture. Dry SS contains on average 50–70% organic matter and 30–50% mineral components [8]. Physicochemical and biological characteristics of agricultural soils, which are amended with the organics-rich SS, can be considerably improved. Particularly, a reduced bulk density leads to an increased soil porosity and soil-air recirculation, as well as improved soil structure and water holding capacity. Besides, the concentration of soil humus is increased. Organic matter of SS enhances soil nutrient storage, soil biota, and diversity, as well as reduces exposure to erosion. High organic matter content facilitates the formation of stable organic complexes with humic acids, thus reducing metal availability [11]. A slow release of mineral elements from SS to soil also changes the physical, chemical, and biological parameters of soil and benefits from increased gas exchange, better water infiltration, and its retention. The compounds of SS are available for a longer period [12].

Mbagwu and Piccolo [13] found that the decomposition of organic materials in sludge enhances the availability of nutrients such as nitrogen and phosphorus substantially. Application of SS at a rate of 200 t ha^{-1} increased the total nitrogen of soil aggregates by 57% and available phosphorus by 64.2%.

The formation of organic and inorganic acids throughout the decomposition process of SS components under aerobic conditions increases soil acidity. Soil salinity positively correlates with the increased application rate of SS. Amendment of loamy-clay soil with SS at dose 60 t ha^{-1} increased soil carbon content from 0.16% to 1.45% [14].

Comparison of physicochemical characteristics of SS of different origins showed that average concentrations of nitrogen, phosphorus, and potassium are similar and reach up to 3.20%, 1.75%, and 0.5% per kg of treated dry SS, respectively [15–17]. Often the potassium content in SS is considered insufficient for plant nutrition [8]. Some studies indicate that SS is an efficient replacement for chemical fertilizers, especially phosphorus. Indeed, Switzerland, Germany, and Austria are developing legislation to make P recovery mandatory from municipal SS [18, 19].

Nevertheless, there are some site-specific factors (e.g., applied technology, quantity, and the origin of raw wastewaters, which differ by the composition of macro- and microelements and risk compounds), that make each SS unique, hence this specificity must be considered to predict the outcome of SS treatment. Determination of these factors remains challenging [20].

3. Legislative aspects in the use of sewage sludge in agriculture

The management of SS in the EU is regulated by various legislative acts. The Directive 2008/98/EC establishes the fundamental ideas and terminologies, such as waste, recycling, and recovery [21]. It explains the basic concepts of waste management, the distinction between waste and secondary raw material ("end-of-waste criteria"), waste, and by-products. The directive lays down basic principles of waste management without adversely affecting human life, health, nature, and the environment. Waste legislation and policy of the EU Member States shall apply as a priority order with the waste management hierarchy (**Figure 1**).

An ex-post evaluation of the SS Directive 86/278/EEC in 2014 showed that its initial objectives were achieved, in spite of large variations in the amount of SS used in agriculture in the Member States (from none to well over 50%) [21, 22]. Currently (2020–2021), the EU initiated an evaluation of legislation efficiency, as well as the risks and opportunities of SS used in farming [23, 24].

Furthermore, two EU working documents on sludge have been produced: the EU Working Document on sludge (2000) and the EU Working Document on sludge and biowaste (2010). The EU Working Document on sludge (2000) indicates that to be used without restrictions, sludge should undergo an hygienization process by an "advanced treatment," which should result in at least a 6-log-unit reduction in *Escherichia coli*, as well as create a sludge that meets the following criteria: In 50 g, there is no Salmonella (wet weight, WW) and *E. coli* <500 colony-forming unit CFU g^{-1}. It was also proposed that sludge produced by "conventional treatments" should show a 2-log-unit reduction of *E. coli*, and its use is allowed with restrictions on its application time, site, and modality. Mesophilic anaerobic digestions at a temperature of 35°C with a mean retention time of 15 days and thermophilic anaerobic digestions at a temperature of at least 53°C for 20 h as a batch, without admixture or withdrawal during the treatment, are indicated, among others, as conventional and advanced treatment processes, respectively. The more recent EU document only suggests the limited absence of Salmonella in 25–50 g and *E. coli* $<5 \times 10^5$ g^{-1} WW as possible criteria for the use of sludge in agriculture [25].

According to the EPA Environmental guidelines published in 2000 on stabilization of biosolids products [26], a biosolids product must meet at least one pathogen reduction requirement and at least one vector attraction reduction requirement [27]. Stabilization Grade A includes thermally treated biosolids (at least 50°C), high pH-high temperature process and biosolids from unknown processes, while stabilization Grade B—anaerobic digestion, aerobic digestion, air drying, composting, lime stabilization, extended aeration, and other processes accepted by the EPA products [26].

Figure 1.
The Waste Framework Directive 2008/98/EC priority order with the waste management hierarchy [21].

4. Economical aspects: technological efficiency and circular economy

In the context of sustainable development principles some main components, which determine the rational solution of the multi-faceted problem of municipal SS, must be considered. Poor farming practices combined with the overuse of chemical fertilizers on poor soils have caused a negative environmental impact, which leads to the degradation of arable land. The effort to increase productivity by increasing the use of various chemicals in fertilizers further diminishes soil fertility. With each harvest, the soil loses organic compounds, and permanent aggravation of improper agricultural practices often prevents the land from recovering. World chemical fertilizer consumption increased from 70.95 kg/ha in 1976 to 138.16 kg/ha in 2016. And in some regions, the fertilization dose increased up to hundreds and even thousands of kilograms per hectare (**Figure 2**) [28].

The quantity of organic elements in the soil constantly decreases. A significant part of the SS does not return to the soil, but is disposed into the sea, is incinerated, or is subject to other different kinds of destructive effects, leading to drastic decreases in soil fertility and continuous soil degradation.

The quantity of the SS constantly increases. The peculiarity of SS lies in its multimineral compound and a huge range of organic matter; in fact, the SS is a nitrogen-phosphorus-potassium organic fertilizer, containing a full set of microelements necessary for the growth of crops. However, due to the high risk of pathogenic impact, a huge part of human and material resources is directed to the destruction of this important resource.

The overwhelming majority of the SS disposal methods are expensive, harmful, or contain both factors. Most municipalities face the growing problem of wastewater treatment. In many cases, waste is dumped into landfills, oceans, or incinerated. The rational solution to the problem of municipal SS disposal lies in an integrated approach to returning the sludge into the agricultural cycle [29].

The directive introduces the "polluter pays" principle and the extended producer responsibility. Some existing projects of producing energy, for example, biogas, minerals, and chemicals out of the sludge, do not prove to be sustainable and viable financially. Furthermore, in most cases, most of the sludge is eventually dumped at the end of the process. Incineration represents the total elimination of the sludge but is extremely expensive. It seems to be the most rational to consider SS not as a problem, but as a valuable resource.

In recent years, out of concern for the profound soil degradation, a growing trend of shifting to organic fertilizers is taking over within the agricultural industry.

The global fertilizer market was valued at around $360 billion before the COVID-19 pandemic with organic fertilizer making up just $6.8 billion. The organic fertilizer market is described as steadily increasing and expected to post a CAGR (Compound Annual Growth Rate) of 14% during the period 2019–2023, with the key factor being increased food demands and agricultural shortages due to population growth and climate change [30].

In case of the continuing negative influence of the high transport, logistics, and energy costs, the SS processing can offset the lack of fertilizers through a domestic product that costs only a fraction of the price to make, creating a local commodity with a considerable economic edge.

Sewage sludge is a natural epidemic focus, and the detection of SARS-CoV-2 in fecal masses led to the long-overdue conclusion to strengthen human health protective measures and counteract the emergence of epidemics [31]. The necessity of the

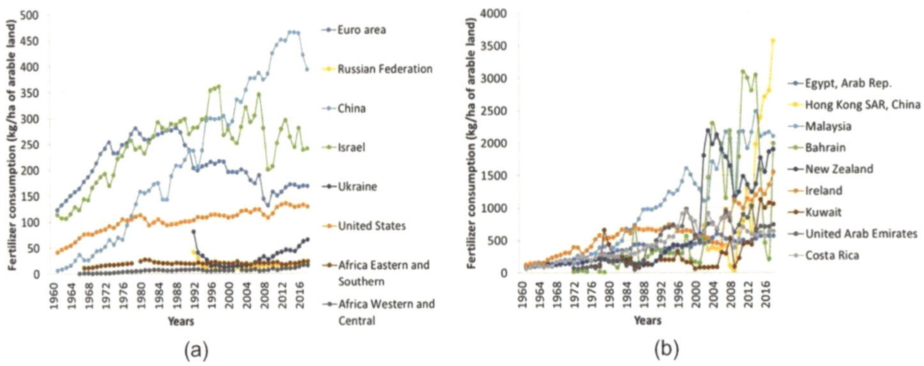

Figure 2.
Fertilizer consumption by different countries and regions. (a) Data on some countries and regions with fertilizer consumption below 500 kg/ha; (b) data on countries with rapid growth of fertilizer consumption, which exceeded 500 kg/ha [28].

implementation of new biological safety criteria can have a significant economical and long-term structural influence on the development of the entire sphere of processing and use of SS. For instance, regarding the sediment formed during the epidemic, it is recommended to avoid its traditional aerobic composting. At once, in the sludge undergoing thermal disinfection treatment, the risk of infection with SARS-CoV-2 is considered in the range from low to negligible [32]. Intensive decontamination measures will make the product more expensive, but more in line with the requirements of sustainable development.

To prevent potential biological threats toward the environment and human health, it becomes increasingly important to develop the most isolated from the environment hermetic methods for the SS disposal, without destroying the organic component, valuable for agriculture.

Economic aspects of SS hygienization have been analyzed [33]. The energy requirement per 100 tons of sludge was estimated depending on different disinfection conceptions. Thus, solar dehydration and chemical treatment with alkali consume 11.7 and 148.3 kW h with the production of 80 tons and 99.6 tons, respectively. In turn, the most expensive technology is gamma irradiation, which consumes 64,800 kW h for obtaining 97.6 tons of the product. The thermal drying also requires quite a high energy consumption, that is, 21,000 kW h for 20 tons of product. The composting does not consume electricity [33]. The high costs of thermal hydrolysis and ultrasonic methods and the need for a neutralizing agent in acid solubilization limit the rapid implementation of these processes in industrial practice [34].

Our testing of the infrared heating method for SS disinfection demonstrated successful results. It took 15 min for the material with an 80% humidity, including the time it required to heat the layer to 95°C, which is below the temperature at which the organic matter decomposes [35].

The widespread usage of SS biomethanation has resulted in the building of a number of complex installations that combine biological wastewater treatment facilities with anaerobic digesters. The development of digestate-derived granulated soil fertilizers is based on physicochemical processing of biostabilized sludges, in keeping with the circular economy concept and the concept of "waste-to-product" [36].

In this respect, the costs of pretreatment technologies for SS biomethanation with further conversion of digestate to fertilizers should be taken into consideration.

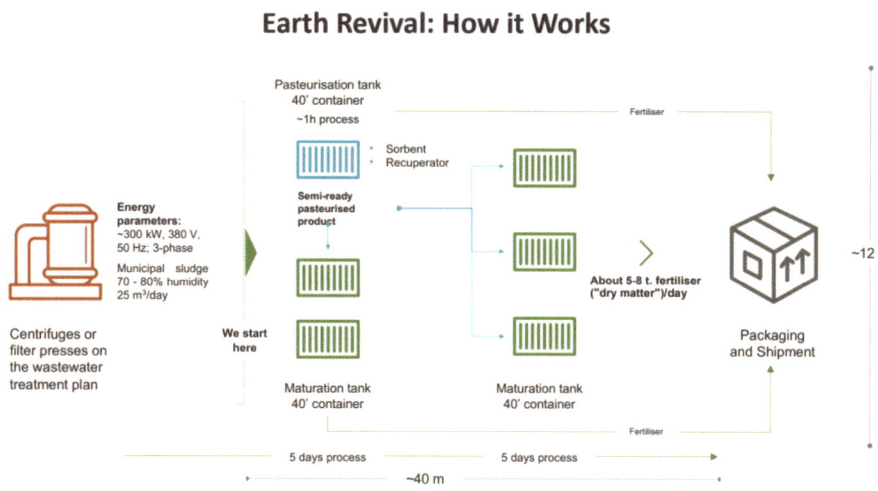

Figure 3.
Technology of the fast recycling of SS into organic fertilizer. Methods are according to Chukurna et al. [43].

The estimated energy utilized for the mechanical operations during SS disintegration and anaerobic digestion (stirring and pumping) was calculated to be 1253.6 kW h per ton [37]. The energy spent for SS pretreatment may vary depending on the solubilization [38], used consumables [39], and methods [40]—thermochemical (TC), sonic, thermo-chemo-sonic, etc. It is experimentally proven that combined disintegration pretreatment should be more efficient. The energy consumption for TC sludge pretreatment (30% solubilization) for biogas production was calculated to be 1588.552 kW h per ton of sludge. The thermos-chemo-ozone (TCO3) pretreatment can optimize the total energy input up to ~721.766 kW h per ton [41].

The evaporation of water should be weighed out between the energy costs in the process and the SS management costs without drying [42].

According to the economic feasibility review of our project for fast SS recycling into biological fertilizers, the energy cost will be nearly $30 per ton of fertilizers (with its humidity ~50% and energy costs $0.1 per kW h and initial SS humidity ~80%). The tested method allows providing 1 ton of bio-pathogenic-free fertilizers due to utilizing up to 1.5 tons of SS and withal avoid other SS disposal costs (**Figure 3**) [43].

The applied methods and technical decisions have international priority under the Paris Convention for the Protection of Industrial Property, the World Intellectual Property Organization (WIPO) Eurasian Patent Organization (EAPO) and national patent organizations.

5. Sewage sludge treatment technologies

5.1 Stabilization

Stabilization of SS aims at reducing some disadvantages of SS (e.g., odor, leaching of heavy metals, etc.), thus considerably extending the potential of SS application. The extent to which readily biodegradable organic matter has degraded is referred to as the degree of stability [44]. Mixing of SS with fly ash, lime, peat, clay, straw, and other residues considerably improve SS characteristics, reducing leachability for metals and soil loss [45, 46]. The addition of wheat straw to the bioaugmented SS after 16 days incubation demonstrated the highest and most stable respiration intensity, the lowest ammonia emission, and the highest stimulation effect on the cress seedling growth, as compared to other treatment types [47].

Santos et al. [22] compared the performance of six residues serving for (i) sludge drying and (ii) improving agronomic properties of the final product. Weathered coal fly ash, bottom biomass ash, green liquor dregs, lime mud, eggshell, and rice husk were chosen as adjuvants based on circular economy and industrial ecological parameters. (0.15 g adjuvant/g SS wet basis). The addition of bottom biomass ash to SS promoted the highest diffusion coefficient and drying rate. The highest positive effect on agronomic parameters was shown for the SS amended with eggshell. Among evaluation criteria were acid neutralization capacity, oxygen uptake rate, and germination index [22].

5.2 Disinfection

Sludge treatment technologies for preparing a valuable fertilizer must meet legislative criteria on sludge hygienization. Numerous technological approaches on SS treatment, which were conducted at ambient temperature or under mesophilic conditions, had a strong effect on biological liquid sludge stabilization and natural dewatering and drying technologies, although disinfection efficiency was unsatisfactory [48–50]. In this respect, further comprehensive research on SS treatment should be focused on a combination of different physical (especially, thermal) and chemical processes, which would convert SS into a qualitative fertilizer with safe microbiological characteristics. **Figure 4** summarizes a broad spectrum of methods for SS disinfection. Several studies have experimented with hybrid methods where two or more technologies can be integrated to increase treatment efficiency and performance [62].

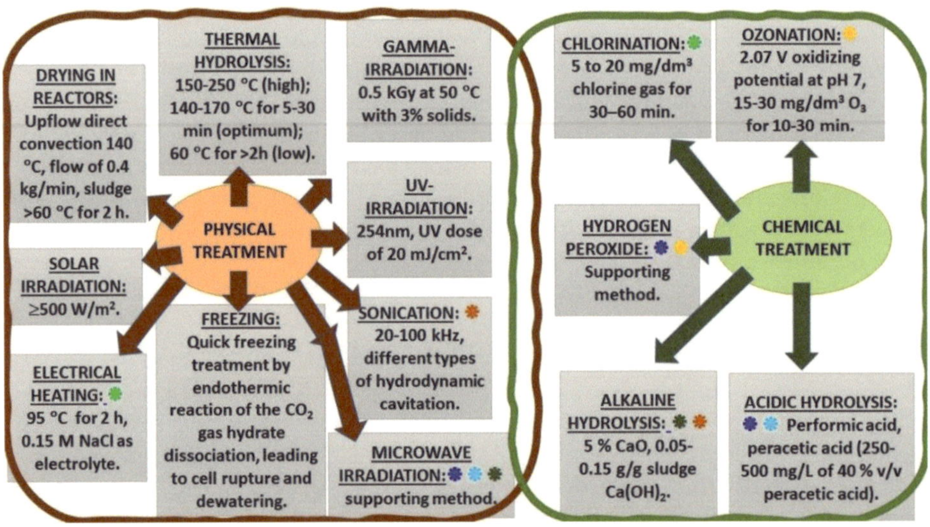

Figure 4.
Methods of sludge disinfection. Combination of different methods is indicated by asterisks of the same color. By Izydorczyk et al. [34, 51–61].

The disinfection approaches should be optimized to minimize potential adverse impacts, such as antimicrobial resistance [62]. Another inherent problem with all sludges rich in nutrients is pathogen regrowth. Offensive odors serve as indicators of microbial regrowth because they are produced as bacteria break down proteins and other organic compounds containing nitrogen and sulfur [63].

6. Changes of microbial community composition in the sewage sludge and soil upon sludge treatment and application

The SS is characterized by a great microbial diversity, which may vary depending on the origin of sewage, its treatment, and industrial activity. Microbial activity in SS, transformation by-products, and residues may impact soil quality if SS is used as fertilizer/amendment [64]. The number of different groups of indicator microorganisms in 1 g of raw SS (wet) on average is 10^2–10^3 for Salmonella (bacteria), Enteroviruses (viruses), Giardia (protozoa), and Ascaris (helminths), while 10^6 – for bacteria *Escherichia coli* [56, 65].

6.1 Microbial community structure in the raw and treated sewage sludge

Many factors modulate microbial community structure within SS, which may change from autotrophic to heterotrophic bacteria depending on the effluent source. According to Nascimento et al. and Nielsen et al. [64, 66], Proteobacteria phylum (21–65%) is predominant in municipal SS. This phylum was primarily dominated by Betaproteobacteria that represents bacteria involved in organic matter degradation and nutrient cycling. Bacteroidetes, Acidobacteria, and Chloroflexi were among the less prominent species. Our recent experiments have also revealed Proteobacteria to dominate in the raw SS (60.17% reads), which consisted of 16.40%, 29.18%, and 12.33% of Alphaproteobacteria, Betaproteobacteria, and Gammaproteobacteria, respectively. At the genus level, the most abundant were Streptomyces (5.68%) and Pseudomonas (3.48%) (**Figure 5A**) [47].

Considerable changes in the microbial community structure of SS occur during biological treatment. Proteobacteria and Bacteroidetes were the most abundant in aerobic and anaerobic conditions, respectively [67, 68]. As reported by Rimkus et al. [47],

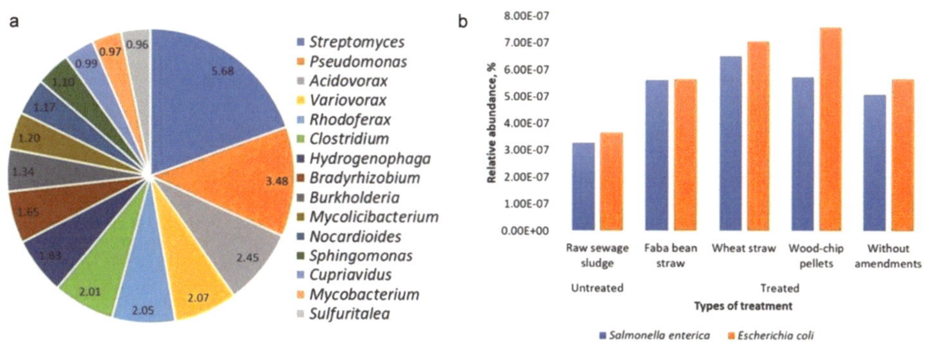

Figure 5.
Relative abundance of microorganisms in sewage sludge. A—at the genus level with relative abundance ≥1% (29% from the total reads); B—Salmonella enterica and Escherichia coli in the sewage sludge before and after the 16-day incubation with different carbon amendments. Methods are according to Rimkus et al. [47].

addition of three types of carbon sources (faba bean straw, wheat straw, and woodchip pellets) to the raw SS resulted in considerable changes in microbial community structure after 16 days of aerobic incubation. In particular, abundance of Firmicutes increased from 5% in the raw SS to 35–50% in the treated samples. In turn, abundance of Proteobacteria decreased from 62% in the raw SS to 32–45% in the treated samples. Yet, the SS incubation without C amendment resulted in a remarkable increase in virus abundance (i.e., 0.34% reads) [47]. The relative abundance of *Salmonella enterica* and *Escherichia coli* has been increased in the treated sludges, as compared to the raw SS (**Figure 5B**).

6.2 Shift in soil microbial community structure after application of the sludge-derived fertilizer

When SS is applied to soil, it causes changes in the structure and functioning of the agroecosystem. The most sensitive component is the soil microbiota, which can undergo both stimulatory and inhibitory changes in the activity and structure. These changes are greatly dependent on soil characteristics and SS application rate.

The microcosm experiment with SS-amended sandy soil (25.71 g SS/kg dry soil) after 119 days has revealed significant changes in prokaryotic community composition at the phylum level, as compared to the non-amended control [48]. Specifically, in SS-amended soil, the relative abundance of Firmicutes reduced from 58.6% at Day 0 to 18.7% at Day 119, while Proteobacteria increased from 15.5% to 36.4%, respectively [69]. In the control soil, these two respective phyla did not change considerably for 119 days. The relative abundance of Actinobacteria in SS-amended soil has increased from 3.1% to 13.2%, while in the control soil decreased from 27.6% to 19.4% [69].

The use of sludge as a soil amendment has been shown to increase the activity of soil enzymes, for example, arylsulfatase, acid phosphatase, and alkaline phosphatase. Basal respiration and the fluorescein diacetate hydrolysis activity increased with increasing the dose of SS [70]. Changes in urease activity by soil microorganisms can be discussed in two aspects. First, urease activity reflects the activity of microorganisms involved in the nitrogen cycle in soil [71]. Another aspect is related to the global loss of nitrogen (up to 70%) due to urease activity if urea is applied as a fertilizer. Therefore, urease inhibition is one of the strategies worldwide to maintain soil fertility [72]. In our experiments, combination of dry SS with nitrogen-containing fertilizer resulted in inhibition of urease activity in loamy soil during the vegetation experiment with maize [73].

The addition of SS-derived organics to soil increases the C_{mic}/C_{total} and N_{mic}/N_{total} ratios in the soil. At the same time, application of SS containing heavy metals, according to Fließbach et al. [8] and Chander and Brookes [74], C_{mic}/C_{total} ratio decreases to 32% and 50%, respectively. This effect can be developed greater in sandy soil than in clayey soils [75].

6.3 Indicators of microbiological contamination

In the early nineteenth century, the total coliforms, fecal coliforms, and fecal streptococci were considered as typical indicator bacteria. Later it was shown that these pathogens are not a major concern in solid waste landfills or leachate [62, 76]. Nowadays, different types of bacteria (fecal coliforms and *Escherichia coli*, *Salmonella*, Shigella, *Vibrio cholerae*); diverse parasite cysts and eggs (*Balantidium coli, Entamoeba histolytica*, and *Giardia lamblia*, helminths); viruses (human adenoviruses, enteroviruses (e.g., polioviruses), diarrhea-causing viruses (e.g., rotavirus), hepatitis-A virus and reoviruses) and fungi are monitored as biological contaminants of SS. Depending on the type and amount, they can all be harmful to the environment and human health [62, 76].

7. Effect of sludge-derived fertilizers on the plant growth

Soil amendment with SS is useful for enhancing crop production, as well as the accumulation of nutrients and organic matter in the soil. However, the accumulation of humic substances (HS) in soil and plant tissues must be regularly observed in case the SS is continuously used [14]. The SS can be used as fertilizers also after pyrolysis [77]. Both sole application of SS and their respective biochars provided enough P for the plants to achieve biomass higher than conventional P-fertilizer [77].

The effect of SS on plant growth differs depending on the SS application method, that is, at the soil surface "mulching" or mixed homogeneously with soil. The application of SS on the surface has some advantages, that is, water evaporation is limited by forming a physical barrier that allows soil moisture to be retained longer. Due to those, the biological and chemical processes of organic matter transformation intensified [78]. For example, the best yield of wheat (*Triticum durum* Desf.) was obtained when SS (dried) is applied at the clayey-silty soil surface (mulching) as compared to homogeneously mixed SS with soil [78]. Plant response to SS in dependence on SS application rate, plant species, soil type, and experiment conditions is shown in **Table 1.**

Importantly, a direct application of SS on agricultural soils is not recommended. It was shown that the hygienically treated (by liming) SS inhibited the growth of white mustard (*Sinapis alba* L.) already at a ratio of 10%. The addition of compost (5%, 15%, and 25%) resulted in the suppressed phytotoxicity of sludge in all tested ratios, that is, from 5 to 50% [92].

Our experiments showed that the use of SS affects the germination and development of seedlings. Concentrations exceeding 7 g kg^{-1} inhibited the germination of cucumber seeds and resulted in necrotizing primary roots. In the study with air-dried SS mixed with agricultural sandy loam soil at rates of 0 (control), 10, 20, 30, 40, and 50 g kg^{-1} (equal to 0, 30, 60, 90, 120, and 150 t ha^{-1}), seed germination of broad beans (*Faba sativa* Bernh.) decreased from 70.0% (control), to 63.3, 56.7, 50.0, 50.0 and 46.7%, respectively [14]. Nevertheless, all the growth and morphometric parameters of broad beans positively respond to SS-amended soil compared to non-amended soil. The most effective for biomass yield of broad beans was the application of 120 t ha^{-1} SS [14]. In experiments with barley, the stimulation effect of SS also was shown, particularly, the addition of SS 40 g kg^{-1} soil led to an increase of dry weight, leaf area, number of leaves, and tillers per plant [18].

Our recent study demonstrated a positive effect of SS on maize growth and soil microbiological activity, when SS is applied in combination with mineral fertilizers [73]. Additional experiments have been performed also with cucumbers and leaf mustard. The SS preparation alone did not provide the plants with mineral nutrients in appropriate values, while the combination of SS preparation with nitrogen-containing fertilizers significantly improved the plant growth and promoted plant development [73] (**Figure 6**). This may have a long-term favorable effect on plant mineral nutrition. Our data also showed that different plants respond to the SS differently.

SS dose/soil	Plant species and effect	Reference
Field experiments		
8.3% w/w/clayey–silty	Yield increase by mulching 65.7%, by mixing 91.5%: wheat (*Triticum durum*)	[78]
15, 30 and 60 t ha^{-1}/ Entisols	Grain yields at the 1st year less than mineral fertilizers by 3–5%, 2nd year increase in average by 13.8%: corn (*Zea mays*)	[12]
20 t ha^{-1}/clayed soils	Grain yield increase by 71–171%: wheat (*Triticum sp.*)	[79]
25 t ha^{-1}/n.d.	Yield increase by 43.5%: radish (*Raphanus sativus*)	[80]
30 t ha^{-1}/n.d.	Yield increase by 26%: sunflower (*Helianthus annuus*)	[81]
40 t ha^{-1}/sandy clay loam	Grain yield increase by 91.6%; 1000-grain weight increase by 26.9%; number of productive tillers by 51.4%: wheat (*Triticum sp.*)	[82]
80 t ha^{-1}/calcareous soils	Head yield increase by 186%: lettuce (*Lactuca sativa*)	[83]
150 t ha^{-1}/n.d.	Increase by 42.3% in comparison with control, decrease by 31.8% in comparison with RDMF, that is, 45.5% and 22.1%; 46.3% and 18.9%; 51.6% and 27.8%; 52.1% and 8.5%; 35.5% and 16.2%: carrots (*Daucus sativus*); turnips (*Brassica rapa*); radish (*R. sativus*); tomatoes (*Solanum lycopersicum*); onion (*Allium cepa*); summer squash (*Cucurbita pepo*), respectively	[84]
250 t ha^{-1}/mudflat saline-alkaline soil	Biomass increase by 399.7% (control 1.3 t ha^{-1}): sweet sorghum (*Sorghum bicolor*)	[85]
Vegetation pot experiments		
40 t ha^{-1}/loamy clay (calcareous)	Increase of DM up to 5%: barley (*Hordeum vulgare*)	[86]
30 t ha^{-1}/alluvial soils	Seed cotton yield (71.4%), lint yield (67.7%), and cottonseed yield (74.1%) were increased: cotton (*Gossypium hirsutum*)	[87]
300 t ha^{-1}/mudflat soil	Fresh weight of aboveground parts and roots increased by 555 and 128%, respectively: ryegrass (*Lolium perenne*)	[88]
15 and 30%, v/v/peat-based medium	Pepper yield and the number of fruits per plant increased by 28–43 and 30–98%, respectively: pepper (*Capsicum annuum*)	[89]
50%, 100%, and 150% of RDMF	Total sugar and sugarcane increased by 4.68 and 4.19%: sugarcane (*Saccharum officinarum*)	[90]
50 and 100%/soil loamy chernozem	Chlorophyll b (15–38%), carotenoids (5–50%) increased, while plant fresh weight (100%) SS was decreased by 8%: Sweet basil (*Ocimum basilicum*)	[91]
8.7 g L^{-1} + MF/loamy sand	Yield increase by 15%: corn (*Zea mays*)	[73]
10, 20, 30, and 40 g kg^{-1}/n.d.	Seed germination rate increased by 9.6, 19.0, 28.6, and 28.6%, respectively, total biomass increased by 146, 236, 278, and 400%: broad bean (*Faba sativa*)	[14]

RDMF: recommended dose of mineral fertilizers; MF: mineral fertilizers.

Table 1.
Plant growth in response to the presence of SS in soil.

Figure 6.
The effect of SSP on the growth of plants: A—cucumbers, B—leaf mustard, C–E—maize. Label color: pink—SSP + NPK, orange—NPK, blue—SSP + PK, green—SSP, yellow—vermicompost, white—soil without fertilizers. SSP—sewage sludge preparation; PK—phosphorus and potassium-containing fertilizer; NPK—nitrogen, phosphorus, and potassium-containing fertilizer. Controls—loamy soil without additional fertilizer, soil mixed with mineral fertilizer (Kristalon 18:18:18). Period of vegetation experiment A—18 days, B—47 days, C—33 days, E—46 days, and D—62 days. The application rate of SSP is 17.3 g L^{-1} in a loamy soil. Methods are according to Dubova et al. [73].

A species-specific effect, in that case, can be explained by (i) different sensitivity of plants to the compounds in SS preparations; (ii) demand for mineral elements at the early stages of ontogenesis due to slow release of nutrients from SS; (iii) insufficient maturing and the presence of growth inhibitors in SS.

8. Environmental impact of the sewage sludge

Sludge production globally in 2017 was 45 MT by dry matter, and now it is increasing annually due to urbanization and population growth [34, 93]. In this respect, the environmental impact of SS in the case of landfill disposal, agricultural use, or other applications is of great importance. Particularly, the contribution of different processes of SS treatment for agricultural use is recently studied by [59]. Energy consumption for SS treatment contributed mostly to global warming (>50%), while SS transportation to agricultural areas affected terrestrial and freshwater ecotoxicity, as well as ozone formation—terrestrial ecosystems (**Figure 7A and B**). Sludge disposal in agricultural areas mostly contributed to human toxicity, terrestrial acidification, and freshwater ecotoxicity (**Figure 7C**). The main impacts of SS in soil are related to the presence of Zn, which affects freshwater ecotoxicity and human toxicity [94].

Biogeochemical emissions from SS handling and spreading on land are expected to be minimized in the future by efficient utilization of nutrients and other resources derived from SS, according to the principles of a circular economy [95, 96].

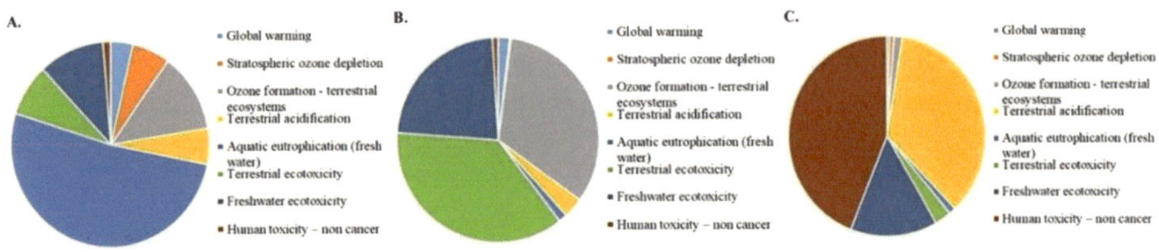

Figure 7.
Environmental life-cycle assessment of the sewage treatment plant: contribution of different activities. A—energy consumption; B—transport of sludge to agricultural areas; C—agricultural areas sludge disposal. By Do Amaral et al. [94].

The processed land-applied SS can emit volatile chemicals and gases that may act alone or in combination with one another to produce the kinds of symptoms [63].

The composition of the sludge and the concentration of pollutants in it predetermine the possibilities of its use. The presence of heavy metals, organic pollutants, and/or pathogens are the main issues associated with the reuse of SS or biosolids extracted from it. According to Manzetti and van der Spoel [97], the following aspects can be reported—(a) raising of the levels of persistent toxins in soil, vegetation, and wildlife, (b) potentially slow and long-termed biodiversity reduction through the fertilizing nutrient pollution operating on the vegetation, (c) greenhouse gas emissions, and (d) the release of odorous compounds. Groundwater contamination from biosolids with pathogenic microorganisms is one of the greatest problems worldwide, due to the lack of adequate and equitable sanitation of SS [98]. Chemical contaminants in processed SS may potentially interact with microbial pathogens, thus, causing or facilitating the disease process via allergic and nonallergic mechanisms, as well as microbial byproducts [63]. Furthermore, endotoxins and exotoxins, which are produced by most bacteria in SS and retain their toxicity at extremely high dilutions, can cause severe illness or death. Endotoxins are heat stable even upon autoclaving, while can be inactivated with dry heat at temperature above 200°C for 1 h [79, 99, 100]. A high microbial diversity of SS leads to the horizontal gene transfer and proliferation of antimicrobial resistance (AMR) [101]. The virus persistence in SS is dependent on the physicochemical and biological properties. For example, enveloped viruses survive for 6–7 days in SS [102], while SARS-CoV-2 might persist on the surfaces up to 72 h [69]. Coronavirus can persist in domestic and hospital SS also for a longer period of time at lower temperatures (4°C) [62, 103].

Long-term accumulation of toxic elements in soil and their uptake by plants is currently the biggest concern in terms of direct SS land application. The bioavailability of heavy metals in the soil is closely related to the value of the soil exchange reaction (soil pH measured in KCl or $CaCl_2$ form), as well as to the sorption properties of the soil, which change with the addition of SS. According to published data, the availability of heavy metals in soils decreases in the order (Zn + Cd) > (Ni + Cu) > (Pb + Cr). However, in connection with physicochemical processes, the accumulation of heavy metals may occur over time, so it is necessary to monitor their concentration for a long time after the application of sludge [104]. When sludge is incorporated into the soil, the heavy metals in it bind to organic matter and clay particles, which usually accumulate in the soil [8, 105].

In Latvia, no more than 14 t ha^{-1} of dry matter may be incorporated at a time with sludge or compost. This corresponds to 55 t ha^{-1} of naturally moist sludge with a dry matter content of about 25% [106]. For 18 years, the concentration of heavy metals in Jelgava SS has significantly decreased. Similar trends have been observed in other treatment plants and this shows that heavy metals are no longer the most important limiting factor for the use of SS.

Wastewater can transport plastics from many different sources, such as fibers from washing machines, personal care products, and facial scrubs. WWTP efficiently removes the microplastics (MPs) from the wastewater, essentially trapping the particles in the sludge [107, 108]. Studies of Peterson [76] showed that 9 years of repeated sludge application led to the accumulation of MPs in the soil. According to various studies, MPs pose various negative effects on soil ecosystems, such as affecting soil fertility, soil organisms' fitness, soil texture, and decreasing crop yield [109, 110].

Pignattelli et al. [111] highlighted the toxicity caused by small MPs (PP, PE, and PVC) on the growth of garden cress (*Lepidium sativum*). Hernández-Arenas et al. [112] studied the effect of MPs in sludge on the growth of tomato plants and discovered that plants grown in soils treated with sludge with a high concentration of MPs had the lowest biomass and did not produce any fruits during the experiment.

Domestic SS is a major source of pharmaceuticals, drugs, and antibiotic resistance genes, so it is important to ensure its biodegradation during sludge treatment. Drugs can remain in the sludge even after stabilization (dewatering), due to their high sorption capacity [113]. Ivanová et al. [114] discovered more than 100 types of drugs and their metabolites in SS. The amount and type of antibiotics in wastewater affect also the composition of bacteria [115].

Pharmaceutical substances are subject to thermal decomposition over a wide temperature range; therefore, it is possible to expect a reduction in the content or their complete removal during thermal processes [116]. Szabová et al. [117] achieved almost 100% drug removal in the sludge by heat treatment at 250°C and incineration at 550°C. Furthermore, pyrolysis at 350–500°C is able to decrease the concentration of MPs in sludge by more than 99% [118].

9. Future direction, challenges, and scope of sewage sludge as a soil fertilizer

The efficient use of waste-derived fertilizers in agriculture needs more empirical knowledge on markets with further research focused on variability, interactivity, and uncertainty. The site-specific factors (e.g., applied technology, quantity, and the origin of raw wastewater differed by the composition of macro- and microelements and risk compounds, soil types, and crops) make each SS unique, hence this specificity must be considered to predict the outcome of SS treatment. New efficient technologies for onsite sludge disinfection are necessary and urgent. Interdisciplinary activities on the safe use of SS upon treatment and application need to be thoroughly analyzed and developed, for example, planning, servicing, diagnosing, storing, and others. Furthermore, macroeconomic factors can considerably influence technology stocks. Soaring gas prices directly affected the production of synthetic fertilizers costs. High prices as well as the disruption of transport and production logistics lead to a real threat of a dramatic reduction in supply on the mineral fertilizers market. Combined with higher prices, an increase in demand for less volatile organic fertilizers can be expected.

10. Conclusions

Summarizing our experimental data on optimization of SS treatment and its application in agriculture, as well as recent findings of other authors in this field, the following conclusions were drawn:

The technology, which was newly developed by Earth Revival Ltd., offers an innovative and comprehensive solution to the problem of SS disposal and soil degradation, which includes aspects of agriculture, healthcare, epidemics, ecology, economics, and the social sphere. Costs can be recuperated through sludge treatment service fees and fertilizer sales.

The infrared heating system, used for SS disinfection, has shown consistently successful results. For the material with a humidity of 80%, it took 15 min, considering the heating time of the layer to 95°C, which is below the temperature of the organic matter decomposition. Research and experiments related to the neutralization of the spore-forming bacteria are planned to be realized in the next stage of the project.

The technology (SS transportation system, maturation process) can be used for fast, safe, and efficient SS processing into organic fertilizer. It also combines well with the anaerobic digestion process as it can complete the digestive sludge transformation to a huge quality fertilizer.

Sewage sludge can replace mineral fertilizers in crop production. Attention should be paid to the amount and ratio of mineral elements available to the plant during plant growth. Sewage sludge may not fully provide plants with potassium and phosphorus. Sewage sludge is recommended for plants with a longer vegetation period due to the slow release of nutrients. A phytotoxic effect may occur during seed germination.

Acknowledgements

This research is being conducted based on agreement with SIA “ETKC” (Centre of Competence for Energy and Transportation) within the framework of project Nr.1.2.1.1/18/A/001 co-funded by the European Regional Development Fund.

Conflict of interest

The authors declare that they have no known competing financial interests or personal relationships that could have appeared to influence the work reported in this paper.

Author details

Olga Muter[1]*, Laila Dubova[2], Oleg Kassien[3], Jana Cakane[2] and Ina Alsina[2]

1 Faculty of Biology, Department of Microbiology and Biotechnology, University of Latvia, Riga, Latvia

2 Institute of Soil and Plant Sciences, Latvia University of Life Sciences and Technologies, Jelgava, Latvia

3 Earth Revival Technologies Ltd., Latvia

*Address all correspondence to: olga.mutere@lu.lv

References

[1] Gray NF. Water Technology: An Introduction for Environmental Scientists and Engineers. 3rd ed. IWA Publishing; 2010. p. 747

[2] Zhang S, Yang Z, Lv X, Zhi S, Wang Y, Li Q, et al. Novel electro-dewatering system for activated sludge biosolids in bench-scale, pilot-scale and industrial-scale applications. Chemical Engineering Research and Design. 2017;**121**:44-56

[3] Buta M, Hubeny J, Zieliński W, Harnisz M, Korzeniewska E. Sewage sludge in agriculture—The effects of selected chemical pollutants and emerging genetic resistance determinants on the quality of soil and crops—A review. Ecotoxicology and Environmental Safety. 2021;**214**:112070

[4] Grubel K, Machnicka A, Nowicka E, Wacławek S. Mesophilic-thermophilic fermentation process of waste activated sludge after hybrid disintegration. Ecological Chemistry and Engineering. 2014;**21**:125-136

[5] Trošanová M, Skultétyova I, Rusnák D. State of the management of municipal waste in the context of reverse logistics. Pollack Period. 2018;**13**:117-128

[6] Duan Y, Pandey A, Zhang Z, Awasthi MK, Bhatia SK, Taherzadeh MJ. Organicsolidwastebiorefinery:Sustainable strategy for emerging circular bioeconomy in China. Industrial Crops and Products. 2020;**153**:112568

[7] Kecskésová S, Imreová Z, Kožárová B, Drtil M. Opätovné získavanie fosforu úpravou čistiarenského kalu. Chemicke Listy. 2020;**114**:341-348

[8] Fließbach A, Martens R, Reber HH. Soil microbial biomass and microbial activity in soils treated with heavy metal contaminated sewage sludge. Soil Biology and Biochemistry. 1994;**26**(9):1201-1205

[9] Collivignarelli MC, Abbà A, Frattarola A, Miino MC, Padovani S, Katsoyiannis I, et al. Legislation for the reuse of biosolids on agricultural land in Europe: Overview. Sustainability (Switzerland). 2019;**11**:6015

[10] Mercl F, Košnář Z, Najmanová J, Hanzlíček T, Száková J, Tlustoš P. Evaluation of mineral nutrient and trace element concentrations in anaerobically stabilized sewage sludge. Waste Forum. 2018;**1**:78-84

[11] Picariello E, Pucci L, Carotenuto M, Libralato G, Lofrano G, Baldantoni D. Compost and sewage sludge for the improvement of soil chemical and biological quality of mediterranean agroecosystems. Sustainability. 2021;**13**:26

[12] Cuevas G, Martínez F, Walter I. Field-grown maize (*Zea mays* L.) with composted sewage sludge. Effects on soil and grain quality. Spanish Journal of Agricultural Research. 2003;**1**:111-119

[13] Mbagwu JSC, Piccolo A. Carbon, nitrogen and phosphorus concentrations in aggregates of organic waste-amended soils. Biological Wastes. 1990;**31**:97-111

[14] Eid EM, Alrumman SA, El-Bebany AF, Fawy KF, Taher MA, Hesham AEL, et al. The evaluation of sewage sludge application as a fertilizer for broad bean (*Faba sativa* Bernh.) crops. Food Energy Security. 2018;**7**:e00142

[15] Aleisa E, Alsulaili A, Almuzaini Y. Recirculating treated sewage sludge for

agricultural use: Life cycle assessment for a circular economy. Waste Management. 2021;**135**:79-89

[16] Wierzbowska J, Sienkiewicz S, Krzebietke S, Sternik P. Sewage sludge as source of nitrogen and phosphorus for Virginia fanpetals. Bulgarian Journal Agricultural Science. 2016;**22**:722-727

[17] Vibornijs V, Rimkus A, Dubova L, Bekkers D, Strunnikova N, Kassien O, et al. Evaluation of sewage sludge for further nutrient conservation. Key Engineering Materials. 2020;**850**:166-171

[18] Eid EM, Alamri SAM, Shaltout KH, Galal TM, Ahmed MT, Brima EI, et al. A sustainable food security approach: Controlled land application of sewage sludge recirculates nutrients to agricultural soils and enhances crop productivity. Food and Energy Security. 2020;**9**:e197

[19] European sustainable phosphorus platform. Scope Newsletter. 2014;**107**:19p

[20] González D, Colón J, Gabriel D, Sánchez A. The effect of the composting time on the gaseous emissions and the compost stability in a full-scale sewage sludge composting plant. Science of the Total Environment. 2019;**654**:311-323

[21] European Commission. Directive 2008/98/EC on Waste Waste Framework Directive. Environment-European Commission; 2008

[22] Santos AF, Vaz TE, Quina MJ. Valorization of industrial wastes with complementary properties for producing organic amendments. In: Proceedings of the 8th International Conference on Sustainable Solid Waste Management. THESSALONIKI. 2021. Available from: http://uest.ntua.gr/thessaloniki2021/pdfs/THESSALONIKI_2021_Santos_et_al%20(3).pdf [Accessed: April 6, 2022]

[23] Nozela W. Caracterização do lodo de esgoto, após desaguamento e secagem térmica. Aleph: da Estação de Tratamento de Esgoto de Araraquara; 2014 Available from: http://acervodigital.unesp.br/handle/11449/124485

[24] European Commission. Sewage sludge use in farming evaluation. 2021. Available from: https://ec.europa.eu/info/law/better-regulation/have-your-say/initiatives/12328-Sewage-sludge-use-in-farming-evaluation_en

[25] Levantesi C, Beimfohr C, Blanch AR, Carducci A, Gianico A, Lucena F, et al. Hygienization performances of innovative sludge treatment solutions to assure safe land spreading. Environmental Science Pollution Research. 2015;**22**:7237-7247

[26] EPA. Environmental Guidelines: Use and Disposal of Biosolids Products. Sydney: Waters & Catchments Policy Section, Environmental Policy Branch, Environment. Protection Authority; 2000

[27] EPA. Technical Support Documents for 40 CFR Part 503. Land Application of Sewage Sludge, No Title; 1993

[28] The World Bank. Fertilizer consumption [Internet]. Available from: https://data.worldbank.org/indicator/AG.CON.FERT.ZS [Accessed: April 6, 2022]

[29] European Commission. Waste and recycling [Internet]. Available from: https://ec.europa.eu/environment/topics/waste-and-recycling_en [Accessed: April 6, 2022]

[30] Businesswire. Global Organic Fertilizers Market 2019-2023|14%

CAGR Projection Over the Next Four Years|Technavio [Internet]. 2019. Available from: https://www.businesswire.com/news/home/20190106005072/en/Global-Organic-Fertilizers-Market-2019-2023-14-CAGR-Projection-Over-the-Next-Four-Years-Technavio

[31] van Doremalen N, Bushmaker T, Morris DH, Holbrook MG, Gamble A, Williamson BN, et al. Aerosol and surface stability of SARS-CoV-2 as compared with SARS-CoV-1. The New England Journal of Medicine. 2020;**382**:1564-1567

[32] ANSES. Sewage sludge produced during the COVID-19 epidemic can only be applied to fields after disinfection [Internet]. 2020. Available from: https://www.anses.fr/en/content/sewage-sludge-produced-during-covid-19-epidemic-can-only-be-applied-fields-after

[33] Diocaretz MC, Vidal G. Technical and economic aspects of the sewage sludge disinfection processes from municipal wastewater treatment. Theoría - Ciencia, Arte y Humanidades. 2010;**19**:51-60

[34] Izydorczyk G, Mikula K, Skrzypczak D, Trzaska K, Moustakas K, Witek-Krowiak A, et al. Agricultural and non-agricultural directions of bio-based sewage sludge valorization by chemical conditioning. Environmental Science and Pollution Research. 2021;**28**:47725-47740

[35] Vaysman V, Strunnikova N, Chukurna O, Dobrovolskyi V, Kassien O. Improvement of the quality of the human environmentbytransportingandstabilizing sewage sludge for further processing. In: Tonkonogyi V, Ivanov V, Trojanowska J, Oborskyi G, Pavlenko I, editors. Advanced Manufacturing Processes III. InterPartner 2021. Lecture Notes in Mechanical Engineering. Springer, Cham; 2022. DOI: 10.1007/978-3-030-91327-4_44

[36] Kaszycki P, Głodniok M, Petryszak P. Toward a bio-based circular economy in organic waste management and wastewater treatment—The polish perspective. New Biotechnology. 2021;**61**:80-89. DOI: 10.1016/j.nbt.2020.11.005

[37] Kavitha S, Yukesh Kannah R, Yeom IT, Do KU, Banu JR. Combined thermo-chemo-sonic disintegration of waste activated sludge for biogas production. Bioresources Technology. 2015;**197**:383-392. DOI: 10.1016/j.biortech.2015.08.131

[38] Banu JR, Kavitha S, Kannah RY, Usman TMM, Kumar G. Application of chemo thermal coupled sonic homogenization of marine macroalgal biomass for energy efficient volatile fatty acid recovery. Bioresource Technology. 2020;**303**:122951. DOI: 10.1016/j.biortech.2020.122951

[39] Rajesh Banu J, Kannah RY, Kavitha S, Gunasekaran M, Kumar G. Novel insights into scalability of biosurfactant combined microwave disintegration of sludge at alkali pH for achieving profitable bioenergy recovery and net profit. Bioresource Technology. 2018;**267**:281-290. DOI: 10.1016/j.biortech.2018.07.046

[40] Kavitha S, Rajesh Banu J, Subitha G, Ushani U, Yeom IT. Impact of thermo-chemo-sonic pretreatment in solubilizing waste activated sludge for biogas production: Energetic analysis and economic assessment. Bioresource Technology. 2016;**219**:479-486. DOI: 10.1016/j.biortech.2016.07.115

[41] Kannah RY, Kavitha S, Rajesh Banu J, Yeom IT, Johnson M. Synergetic effect

of combined pretreatment for energy efficient biogas generation. Bioresource Technology. 2017;**232**:235-246. DOI: 10.1016/j.biortech.2017.02.042

[42] Santos AF, Gomes LA, Góis JC, Quina MJ. Improvement of thermal dehydration and agronomic properties of products obtained by combining sewage sludge with industrial residues. Waste and Biomass Valorization. 2021;**12**:5087-5097

[43] Chukurna O, Vaysman V, Kassien O, Dobrovolskyi V, Strunnikova N, Nitsenko V. Recycling of municipal sewage sludge in sustainable logistics systems. Journal of Information Technology Management. 2022. Accepted

[44] Bożym M, Siemiątkowski G. Characterization of composted sewage sludge during the maturation process: A pilot scale study. Environmental Science and Pollution Research. 2018;**25**:34332-34342

[45] Ros M, Hernandez MT, García C. Bioremediation of soil degraded by sewage sludge: Effects on soil properties and erosion losses. Environmental Management. 2003;**31**:741-747

[46] Zupančič M, Bukovec P, Milačič R, Ščančar J. Critical evaluation of the use of the hydroxyapatite as a stabilizing agent to reduce the mobility of Zn and Ni in sewage sludge amended soils. Waste Management. 2006;**26**:1392-1399

[47] Rimkus A, Gudrā D, Dubova L, Fridmanis D, Alsiņa I, Muter O. Stimulation of sewage sludge treatment by carbon sources and bioaugmentation with a sludge-derived microbial consortium. Science of the Total Environment. 2021;**783**:146989

[48] Metcalf W, Eddy C. Wastewater Engineering: Treatment and Reuse. New York, NY: McGraw Hill; 2013. pp. 2048

[49] Bauerfeld K. Effect of ambient temperatures on disinfection efficiency of various sludge treatment technologies. Water Science & Technology. 2014;**69**:15-24

[50] Junga P, Mach P, Mareek J. Evaluation of efficiency of technologies for wastewater sludge hygienisation. Research in Agricultural Engineering. 2017;**63**:54-61

[51] Pilli S, Yan S, Tyagi RD, Surampalli RY. Thermal pretreatment of sewage sludge to enhance anaerobic digestion: A review. Critical Reviews in Environmental Science and Technology. 2015;**45**:669-702

[52] Sapkaite I, Barrado E, Fdz-Polanco F, Pérez-Elvira SI. Optimization of a thermal hydrolysis process for sludge pre-treatment. Journal of Environmental Management. 2017;**192**:25-30

[53] Huang X, Qu Y, Cid CA, Finke C, Hoffmann MR, Lim K, et al. Electrochemical disinfection of toilet wastewater using wastewater electrolysis cell. Water Research. 2016;**92**:164-172. DOI: 10.1016/j.watres.2016.01.040

[54] Yin Z, Hoffmann M, Jiang S. Sludge disinfection using electrical thermal treatment: The role of ohmic heating. Science of the Total Environment. 2018;**615**:262-271

[55] Fogolari O, Magri ME, Philippi LS. Sanitisation of sewage sludge in a solar heated reactor: Inactivation of total coliforms and *Escherichia coli*. Engenharia Sanitaria e Ambiental. 2018;**23**:91-100

[56] Hawrylik E. Methods used in disinfections of wastewater and sewage sludge—Short review. Architecture, Civil Engineering, Environment. 2020;**13**:57-63

[57] Nowicka E, MacHnicka A. Hygienization of surplus activated sludge by dry ice. Ecological Chemistry and Engineering Science. 2015;**21**:651-660

[58] Kim W, Lee HK, Kwon YN. Investigation of a gas hydrate dissociation-energy-based quick-freezing treatment for sludge cell lysis and dewatering. International Journal of Environmental Research and Public Health. 2019;**16**:3611

[59] Melmed LN, Comninos DK. Disinfection of sewage sludge with gamma radiation. Water SA. 1979;5:153-159

[60] Dietrich JP, Loge FJ, Ginn TR, Başağaoğlu H. Inactivation of particle-associated microorganisms in wastewater disinfection: Modeling of ozone and chlorine reactive diffusive transport in polydispersed suspensions. Water Research. 2007;**41**:2189-2201

[61] Porat I, Brosseau C, Snyder C, Hill C, Stammegna M, Beaudry S. Wastewater disinfection—The smart way. In: WEFTEC 2019—92nd Annual Water Environment Federation's Technical Exhibition and Conference. 2019;**Code 152954**:4128-4146

[62] Anand U, Li X, Sunita K, Lokhandwala S, Gautam P, Suresh S, et al. SARS-CoV-2 and other pathogens in municipal wastewater, landfill leachate, and solid waste: A review about virus surveillance, infectivity, and inactivation. Environmental Research. 2022;**203**:111839

[63] Gattie DK, Lewis DL. A high-level disinfection standard for land-applied sewage sludges (biosolids). Environmental Health Perspectives. 2004;**112**:126-131. DOI: 10.1289/ehp.6207

[64] Nascimento AL, Souza AJ, Andrade PAM, Andreote FD, Coscione AR, Oliveira FC, et al. Sewage sludge microbial structures and relations to their sources, treatments, and chemical attributes. Frontiers in Microbiology. 2018;**9**:1462

[65] Davis RD, Carrington EG, Gendebien A, Aitken MN, Fenlon D, Svoboda I. A users' guide to research on apllication of organic waste to land. In: Report SR 4624/3. 2013

[66] Nielsen PH, Mielczarek AT, Kragelund C, Nielsen JL, Saunders AM, Kong Y, et al. A conceptual ecosystem model of microbial communities in enhanced biological phosphorus removal plants. Water Research. 2010;**44**:5070-5088

[67] Liu H, Luo GQ, Hu HY, Zhang Q, Yang JK, Yao H. Emission characteristics of nitrogen- and sulfur-containing odorous compounds during different sewage sludge chemical conditioning processes. Journal of Hazardous Materials. 2012;**235-236**:298-306

[68] Hu M, Wang X, Wen X, Xia Y. Microbial community structures in different wastewater treatment plants as revealed by 454-pyrosequencing analysis. Bioresource Technology. 2012;**117**:72-79

[69] Wolters B, Fornefeld E, Jechalke S, Su JQ, Zhu YG, Sørensen SJ, et al. Soil amendment with sewage sludge affects soil prokaryotic community composition, mobilome and resistome. FEMS Microbiology Ecology. 2018;**95**:fiy193

[70] Vieira RF, Pazianotto RAA. Microbial activities in soil cultivated with corn and amended with sewage sludge. Springerplus. 2016;**5**:1844

[71] Marschner P, Kandeler E, Marschner B. Structure and function of

the soil microbial community in a long-term fertilizer experiment. Soil Biology and Biochemistry. 2003;**35**:453-461

[72] Modolo LV, da-Silva CJ, Brandão DS, Chaves IS. A minireview on what we have learned about urease inhibitors of agricultural interest since mid-2000s. Journal of Advanced Research. 2018;**13**:29-37

[73] Dubova L, Cielava N, Vibornijs V, Rimkus A, Alsiņa I, Muter O, et al. Evaluation of suitability of treated sewage sludge for maize cultivation. Key Engineering Materials. 2020;**850**:159-165

[74] Chander K, Brookes PC. Residual effects of zinc, copper and nickel in sewage sludge on microbial biomass in a sandy loam. Soil Biology and Biochemistry. 1993;**25**:1231-1239

[75] Khan M, Scullion J. Effect of soil on microbial responses to metal contamination. Environmental Pollution. 2000;**110**:115-125

[76] Peterson ML. Soiled disposable diapers: A potential source of viruses. American Journal of Public Health. 1974;**64**:912-914

[77] Rehman RA, Rizwan M, Qayyum MF, Ali S, Zia-ur-Rehman M, Zafar-ul-Hye M, et al. Efficiency of various sewage sludges and their biochars in improving selected soil properties and growth of wheat (*Triticum aestivum*). Journal of Environmental Management. 2018;**223**:607-613

[78] Boudjabi S, Chenchouni H. On the sustainability of land applications of sewage sludge: How to apply the sewage biosolid in order to improve soil fertility and increase crop yield? Chemosphere. 2021;**282**:131122

[79] Özyazici M. Effects of sewage sludge on the yield of plants in the rotation system of wheat-white head cabbage-tomato. Eurasian Journal of Soil Science. 2013;**2**:35-44. DOI: 10.18393/ejss.00995

[80] Lima VN, Silva RVTO, Nunes P, da Silva PH, Morant K, Andrade RFS, et al. The cumulative effects of sewage sludge compost on *Raphanus sativus* L.: Growth and soil properties. Green and Sustainable Chemistry. 2016;**06**:1-10. DOI: 10.4236/gsc.2016.61001

[81] Albuquerque HC, Zuba Junio GR, Sampaio RA, Fernandes LA, Zonta E, Barbosa CF. Yield and nutrition of sunflower fertilized with sewage sludge. Revista Brasileira Engenharia Agricultura Ambiental. 2015;**19**:533-559. DOI: 10.1590/1807-1929/agriambi.v19n6p553-559

[82] Jamil Khan M, Qasim M, Umar M. Utilization of sewage sludge as organic fertilizer in sustainable agriculture. Journal of Applied Science. 2006;**6**:531-535. DOI: 10.3923/jas.2006.531.535

[83] Sönmez F, Bozkurt MA. Lettuce grown on calcareous soils benefit from sewage sludge. Acta Agriculturae Scandinavica. Section B Soil & Plant Science. 2006;**56**:17-24. DOI: 10.1080/09064710510005813

[84] Usman K, Khan S, Ghulam S, Khan MU, Khan N, Khan MA, et al. Sewage sludge: An important biological resource for sustainable agriculture and its environmental implications. American Journal of Plant Sciences. 2012;**3**:25784. DOI: 10.4236/ajps.2012.312209

[85] Zuo W, Gu C, Zhang W, Xu K, Wang Y, Bai Y, et al. Sewage sludge amendment improved soil properties and sweet sorghum yield and quality in a newly reclaimed mudflat land. Science of the Total Environment. 2019;**654**:541-549. DOI: 10.1016/j.scitotenv.2018.11.127

[86] Ahmed H, Fawy H, Abdel-Hady E. Study of sewage sludge use in agriculture and its effect on plant and soil. Agriculture and Biology Journal of North America. 2010;**1**:1044-1049. DOI: 10.5251/abjna.2010.1.5.1044.1049

[87] Tepecik M, Ongun AR, Kayikcioglu HH, Delibacak S, Elmaci OL, Celen AE, et al. Change in cotton plant quality in response to application of anaerobically digested sewage sludge. Saudi Journal of Biological Sciences. 2022;**29**:615-621. DOI: 10.1016/j.sjbs.2021.09.016

[88] Gu C, Bai Y, Tao T, Chen G, Shan Y. Effect of sewage sludge amendment on heavy metal uptake and yield of ryegrass seedling in a mudflat soil. Journal of Environmental Quality. 2013;**42**:421-428. DOI: 10.2134/jeq2012.0311

[89] Pascual I, Azcona I, Aguirreolea J, Morales F, Corpas FJ, Palma JM, et al. Growth, yield, and fruit quality of pepper plants amended with two sanitized sewage sludges. Journal of Agricultural and Food Chemistry. 2010;**58**:6951-6959. DOI: 10.1021/jf100282f

[90] Gonçalves CA, de Camargo R, Thiago Xavier de Sousa R, Silva Soares N, Camargos de Oliveira R, Cristina Stanger M, et al. Chemical and technological attributes of sugarcane as functions of organomineral fertilizer based on filter cake or sewage sludge as organic matter sources. PLoS One. 2021;**16**:e0236852. DOI: 10.1371/journal.pone.0236852

[91] Burducea M, Lobiuc A, Asandulesa M, Zaltariov MF, Burducea I, Popescu SM, et al. Effects of sewage sludge amendments on the growth and physiology of sweet basil. Agronomy. 2019;**9**:548. DOI: 10.3390/agronomy9090548

[92] Šindelá O, Adamcová D, Zloch J, Vaverková MD. Phytotoxicity of sewage sludge from selected wastewater treatment plants—New opportunities in sewage sludge treatment. International Journal of Recycling Organic Waste Agriculture. 2020;**9**:75-83

[93] Gao D, Li X, Liu H. Source, occurrence, migration and potential environmental risk of microplastics in sewage sludge and during sludge amendment to soil. Science of the Total Environment. 2020;**742**:140355

[94] Do Amaral KGC, Rietow JC, Aisse MM. Evaluation of the environmental life cycle of an stp that employs a low-rate trickling filter as post-treatment of a uasb reactor and different sludge-management alternatives. Revista Ambiente & Agua. 2021;**16**:e2648

[95] Johansson K, Perzon M, Fröling M, Mossakowska A, Svanström M. Sewage sludge handling with phosphorus utilization—Life cycle assessment of four alternatives. Journal of Cleaner Production. 2008;**16**:135-151

[96] Svanström M, Heimersson S, Peters G, Harder R, I'Ons D, Finnson A, et al. Life cycle assessment of sludge management with phosphorus utilization and improved hygienisation in Sweden. Water Science Technology. 2017;**75**:2013-2024

[97] Manzetti S, van der Spoel D. Impact of sludge deposition on biodiversity. Ecotoxicology. 2015;**24**(9):1799-1814. DOI: 10.1007/s10646-015-1530-9

[98] Ranjan R, Kumar L, Sabumon PC. Process performance and reuse potential of a decentralized wastewater treatment system. Water Science Technology. 2019;**80**:2079-2090

[99] Aghalari Z, Dahms HU, Sillanpää M, Sosa-Hernandez JE, Parra-Saldívar R.

Effectiveness of wastewater treatment systems in removing microbial agents: A systematic review. Globalization and Health. 2020;**16**:13

[100] Williams KL. Nonendotoxin microbial pyrogens: Lesser endotoxins and superantigens. In: Endotoxins: Pyrogens, LAL Testing and Depyrogenation. Third ed. 2007. p. 20

[101] Calderón-Franco D, Apoorva S, Medema G, van Loosdrecht MCM, Weissbrodt DG. Upgrading residues from wastewater and drinking water treatment plants as low-cost adsorbents to remove extracellular DNA and microorganisms carrying antibiotic resistance genes from treated effluents. Science of the Total Environment. 2021;**778**:-146364

[102] Casanova L, Rutala WA, Weber DJ, Sobsey MD. Survival of surrogate coronaviruses in water. Water Research. 2009;**43**:1893-1898

[103] Wang XW, Li JS, Jin M, Zhen B, Kong QX, Song N, et al. Study on the resistance of severe acute respiratory syndrome-associated coronavirus. Journal of Virology Methods. 2005;**126**:171-177

[104] Ďuricová A, Samešová D. Distribution of the toxic metals in system water—Sludge in the biological water treatment plant. International Multidisciplinary Scientific GeoConference Surveying Geology and Mining Ecology Management, SGEM. 2014;**2**:185-192

[105] Kandeler E, Tscherko D, Bruce KD, Stemmer M, Hobbs PJ, Bardgett RD, et al. Structure and function of the soil microbial community in microhabitats of a heavy metal polluted soil. Biology and Fertility of Soils. 2000;**32**:390-400

[106] Gemste I, Vucāns A. Notekūdeņu dūņas un to izmantošana (In Latvian). Jelgava; 2002. p. 173

[107] Corradini F, Meza P, Eguiluz R, Casado F, Huerta-Lwanga E, Geissen V. Evidence of microplastic accumulation in agricultural soils from sewage sludge disposal. Science of the Total Environment. 2019;**671**:411-420

[108] Yang J, Li L, Li R, Xu L, Shen Y, Li S, et al. Microplastics in an agricultural soil following repeated application of three types of sewage sludge: A field study. Environmental Pollution. 2021;**289**:117943

[109] Cao D, Wang X, Luo X, Liu G, Zheng H. Effects of polystyrene microplastics on the fitness of earthworms in an agricultural soil. In: IOP Conference Series: Earth and Environmental Science. Vol. 61. 2017. p. 012148

[110] Gao Q, Xu J, Bu XH. Recent advances about metal–organic frameworks in the removal of pollutants from wastewater. Coordination Chemistry Reviews. 2019;**378**:17-31

[111] Pignattelli S, Broccoli A, Renzi M. Physiological responses of garden cress (*L. sativum*) to different types of microplastics. Science of the Total Environment. 2020;**727**:138609

[112] Hernández-Arenas R, Beltrán-Sanahuja A, Navarro-Quirant P, Sanz-Lazaro C. The effect of sewage sludge containing microplastics on growth and fruit development of tomato plants. Environmental Pollution. 2021;**268**:115779

[113] Plekancová M, Szabová P, Šefčíková T, Bodik I, Grabic R, Staňová AV. Čo sa deje s farmaceutikami v tepelne upravenom kale? In: 29 konferencia KALY A ODPADY [Internet]. Asociácia čistiarenských expertov Slovenskej republiky; 2020. pp. 87-90. Available from: https://acesr.sk/

wp-content/uploads/2020/10/Zbornik-KaO2020.pdf

[114] Ivanová L, Mackuľak T, Grabic R, Golovko O, Koba O, Staňová AV, et al. Pharmaceuticals and illicit drugs—A new threat to the application of sewage sludge in agriculture. Science of the Total Environment. 2018;**634**:606-615

[115] Błaszczyk K, Krzyśko-Łupicka T. Microbial diversity of sewage sludge. In: Proceedings of ECOpole. 2013. pp. 461-466

[116] Bodík I, Mackuľak T, Fáberová M, Ivanová L. Occurrence of illicit drugs and selected pharmaceuticals in Slovak municipal wastewater. Environmental Science and Pollution Research. 2016;**23**:21098-21105

[117] Szabová P, Varjúová D, Grabic R, Staňová A, Bodík I. Vplyv tepelnej úpravy na obsah farmaceutík v kaloch. In: Časopis Sovak č. 2021. p. 20

[118] Ni BJ, Zhu ZR, Li WH, Yan X, Wei W, Xu Q, et al. Microplastics mitigation in sewage sludge through pyrolysis: The role of pyrolysis temperature. Environmental Science & Technology Letters. 2020;7:961-967. DOI: 10.1021/acs.estlett.0c00740

Chapter 7

Evaluating Waste-to-Energy Technologies as a Waste Management Solution for Uganda

Charlene Nagawa

Abstract

Currently, the world generates 2.01 billion tonnes of waste annually and this is expected to increase to 3.401 billion tonnes of waste by 2050. The continual generation of waste is at the forefront of combating climate change because the waste generated is associated with GHG emissions among other environmental concern. Literature reports that developing countries are lagging the developed countries in waste management and yet these regions are expected to account for most waste generated by 2050. This chapter focuses on the application of Waste-to-Energy (WTE) Techniques in Uganda (developing country) as a way of managing waste, and recommends policies that the Government of Uganda could adopt from the UK to successfully implement these initiatives. The WTE technologies analysed are landfill gas recovery, anaerobic digestion, incineration, pyrolysis, and gasification. The chapter also reviews the current solid waste situation in Uganda with a comparative analysis of the technologies. Since Uganda is a low-income country, it is advised that the country enters Public-Private Partnerships where the developers build and own the technologies. The assessment is informed by literature and personal judgement. Recommendations are made to the GOU on how best to support stakeholders of WTE initiatives further areas of study are highlighted.

Keywords: waste management, waste management hierarchy, waste-to-energy technologies, technology applicability, policy adoption, research

1. Introduction

According to the World Bank, the world generates 2.01 billion tonnes of MSW every year and is expected to increase by 70% by 2050 [1]. The mean global temperature increased by 1.9°F since 1880 and is credited to an increase in human activities like waste generation [2]. Specifically, food waste is a global concern with up to 1/3rd of food produced being wasted in high-income countries and this makes it the world's third-largest emitter [3]. The global population, urbanisation, and economic growth are identified to have a strong correlation to how much waste is produced and yet by 2050, the world population is anticipated to increase to more than 9 billion people with developing economies accounting for 85% of the global population [4] and with the number of people living in urban areas increasing to 6 billion [5].

WTE technologies, through converting the waste generated to energy, are a partial renewable energy source because of the waste that comprises of biomass material like paper, card, and timber. Even though these biodegradable wastes might emit carbon dioxide when burned, the process of photosynthesis enables the plants to absorb the CO_2 and this is regarded as a short-term cycle. Note that the waste which is fossil fuel-based generates GHG and contributes to climate changes [6, 7].

Developing countries are failing to invest in waste management and the cry is to take immediate action to reduce this or the effects of waste on the environment through waste management techniques [8]. Also, developed countries are currently promoting economic and social wellbeing through energy supply yet the energy needs of developing countries are still straining to the respective governments [5]. The success of WTE in Europe is observed by determined investors who trust the technology could work in Africa [9] hence there is a need to exploit the possibility of applying sustainable WTE technologies to meet the growing energy needs while improving the SWM system.

2. Waste management hierarchy

The waste management hierarchy designates the best solutions for managing waste according to what is most suitable for environment [10] and is presented in **Figure 1**.

2.1 Prevention

Priority is given to preventing waste and this entails using less material, waste reduction at source, or retaining products for long [10]. Prevention of waste is advantageous from any waste management strategy such as energy recovery, recycle,

Figure 1.
Illustration of the waste hierarchy [10].

and landfill because the production of material that becomes waste as well as its treatment is circumvented. The definition of waste prevention includes avoidance of waste creating products, waste reduction at source, increasing the life cycle of a product, and reuse [11]. According to the World Bank Report, the fastest way to manage and decrease waste is to minimise economic activity because when countries urbanise, their economic wealth grows along with standards of living, disposable incomes, and consumption of goods and services which leads to an increase waste generated [12]. Another suggested solution is moving from a linear to circular economy that is by curtailing resource extraction and material inputs, and improving efficiency through developmental designs, recollection, and recycling [8].

2.2 Reuse and recycle

Under these stages, the waste producer is required to check, clean, repair, and reuse the material or shift the use of the material to another function. Useless waste is converted into useful materials, and hazardous waste is turned in harmless material hence improving utilisation [10, 11]. Resource recycling promotes economic, social, and environmental benefits because the country saves on natural resources, decreases energy consumption, promotes employment while decreasing waste and pollution. Useless waste is converting into useful materials, and hazardous waste is turned in harmless material hence improving utilisation [11]. An increase in recyclable material in the composition of waste requires reuse and recycle management, while an increase in organic waste or other unrecyclable material would require other management techniques [13].

2.3 Recovery

Recovery can be through energy recovery techniques and/or using waste for either agricultural purposes or backfilling [10]. Section 3 provides more insight on energy recovery techniques. It involves recovering usefulness from the waste through energy recovery techniques such as anaerobic digestion, fermentation, incineration, gasification and pyrolysis to produce energy (fuel, heat and power), and using waste for backfilling [10]. The recovery of biogas and heat energy from landfills and incineration plants, respectively, will reduce waste generation and help in the appropriate reutilization of resources. Resource utilisation of livestock and manure, agricultural waste, domestic sewage sludge, and other organic SWs during aerobic composting and anaerobic digestion and then recycling the organic substances and nutrients, etc. are some of the efficient ways to realise SW resources and materials recovery systems [11].

2.4 Disposal

Depending on a country's policies, waste is disposed of through grinding, milling, open dumping, landfilling, and compaction or burned in an incinerator without energy recovery. The waste can also be disposed of in other countries only when it has a market in those specific countries [8, 10, 11].

High income countries mostly dispose through landfilling and thermal treatments while middle- and low-countries mostly dispose by open dumping and poorly managed landfilling. However, the middle-income countries operate with managed dumping processes [12].

3. Energy recovery technologies

In 2016, the total investment in biomass and WTE technologies was 6.8 billion USD which was a decline from 19.9 billion USD in 2011, 14.9 billion in 2012, 12.4 billion in 2013, 10.8 billion in 2014 but an increase from 6.7 billion in 2015 [14]. Despite the increase, it is evident that interest in WTE is not growing across the globe and yet it is successful in European countries. Energy from waste can either be heat, power, or a combination of heat and power and/or secondary energy carriers of gas, liquid or solid. The choice is usually dependent on the energy requirements of the country or region [6].

3.1 Landfill with gas recovery

Landfills are semi-natural terrestrial ecosystems remodelled on lands that were formerly used for disposing of waste. Landfills exist in various regions and are commonly defined by their age, the composition of waste, design, and ecological operation. They are usually disposal for MSW and sometimes for hazardous solid wastes when they are secure [15]. The landfills are designed to make sure the waste is separate from the surrounding environment [16]. Also, two design structures are feasible for a landfill that is landfilling (where waste in packed in an unwanted hole) or land raising (where waste is directly dumped on the ground) [17]. The average landfill occupies 600 acres [18].

The by-products of landfills are landfill leachate produced when rainwater penetrates and channels through the decaying waste, and landfill gas produced through bacterial degradation under anaerobic conditions [15]. These products can be hazardous in the following ways. Firstly, when acids from degrading waste mix with other components waste, it could cause the leachate to become toxic. Secondly, landfill gas is a source of GHG emissions comprising methane and it is highly flammable its leakage poses a risk of explosions to the surrounding environment [17]. Since the by-products can escape or diffuse through cracks in the deposited material, landfills are designed to minimise their movement to protect the environment [15, 19]. Liners and a leachate collection system are installed to prevent leachate from moving to surrounding soil and water while a gas collection system or a landfill cap is installed to hinder the gas from escaping to the air. The system consists of vertical or horizontal wells used to access the gas which can be collected for 7–10 years. Also, its average efficiency is reported to be 70–85% [19]. **Figure 2** shows a typical landfill gas system.

The landfill gas produced contains 45–55% methane and is collected through a system of gas pipes and through combustion it produces electric power by running a gas engine and/or turbine. Also, the gas can be used for cooking in nearby communities, or boiler fuel for district heating and industrial purposes and this is demonstrated in **Figure 3** [6, 19]. The energy potential of landfills across regions ranges from 5 to 40 L/kg of waste depending on the organic composition and has a CV of about 4500 kcal/m^3. Also, it is key that the landfill gas is purified to remove any hazardous chemicals [16, 19]. **Figure 4** shows an example of a landfill recovery site in the UK managed by Viridor. Viridor operates 32 landfill sites in the UK which generate a total of 86 MW of power to supply 50,000 homes with power all year [23].

When landfills reach the maximum capacity, they are closed for replenishment through appropriate engineering designs and the older type is normally deserted [15]. Many disposal sites are poorly operated and stay as open dumps which pose a risk to the environment both in the short- and long-term [13]. Landfills must be closely monitored by the respective municipalities to prevent leakage of the by-products [15].

Figure 2.
Schematic of landfill treatment set-up [20].

Figure 3.
Set-up showing the use of gas recovered from a landfill [21].

3.2 Anaerobic digestion

Anaerobic digestion involves a sequence of biological processes through which biodegradable waste is digested by microorganisms in the absence of free oxygen to generate biogas [6]. The most suitable waste for this technology is organic waste that includes food and agricultural wastes like animal slurries and that the main end product of anaerobic digestion is biogas usually containing 60 and 40% of methane and carbon dioxide, respectively [24]. The composition of waste is comparable to

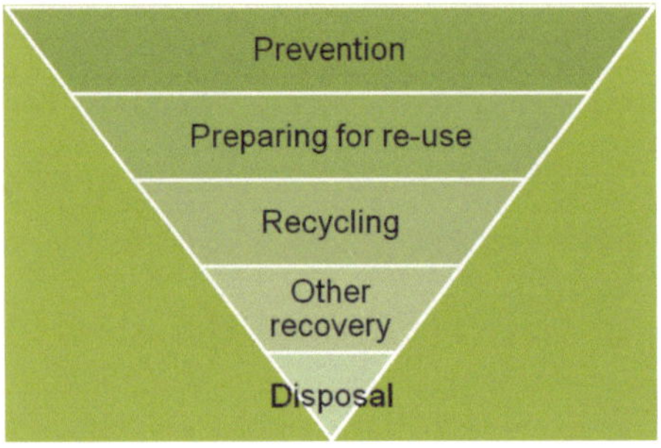

Figure 4.
Viridor Dunbar, a potential landfill energy recovery facility [22].

what is reported by [25] that is 55–70% for methane and 30–35% for carbon dioxide. Anaerobic digestion has relatively long digestion days ranging from 20 to 40 days [24] and because of this decomposition needs to happen by the action of an enzyme [11]. Other factors that affect the process are pH, temperature (35–38°C), loading rate, mixing rate, and toxic compounds [19]. When food waste is added to the process, the methane quantities will increase, and the process of methane production will speed up [24]. The typical amount of biogas produced usually ranges from 50 to 150 m^3/tonne of wastes and this also depends on the composition of the waste [19]. The plants emit residue gases which comprise nitrous oxides, hydrogen chloride, carbon monoxide, and the total organic carbon [25].

The biogas is used to produce electricity and/or heat with a biogas CHP gas engine. Also, the biogas can be used as a renewable natural gas or in the transportation sector as a fuel. The other product of the process is a nutrient-enriched digestate used as a soil fertiliser [19, 24]. **Figure 5** presents a typical illustration of a biogas plant.

This technology is considered to be environmentally friendly and solves the problem of disposing of the bio-degradable waste [11]. Also, anaerobic digestion is usually used to pre-treat the organic component of waste to reduce its weight, and reduce the

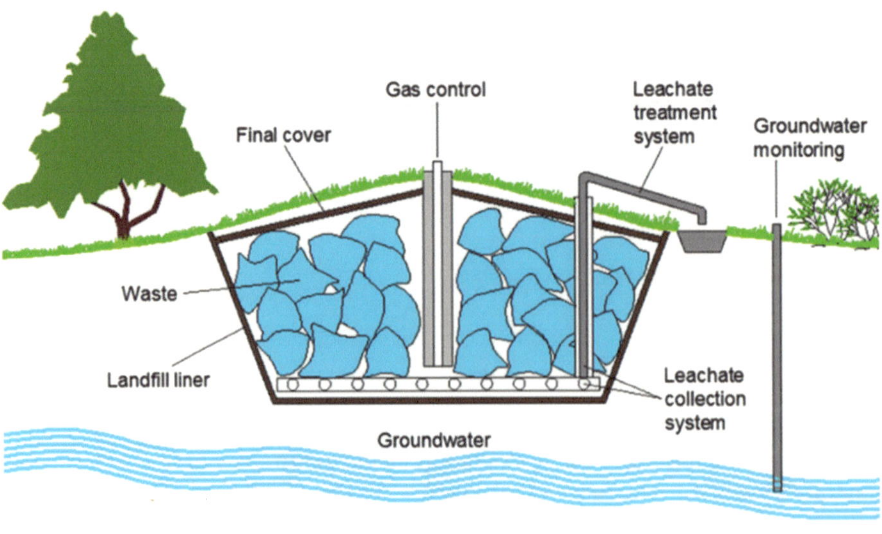

Figure 5.
Schematic of biogas plant [26].

methane and leachate emissions [27]. **Figure 6** shows a biogas plant established in Nakuru, Kenya which uses the crop residues of a farm to generate 2.2 MW of electricity used to cultivate 1740 acres of vegetables and flowers, supplies electricity to up to 6000 rural homes and sells surplus power to the Kenya National Grid [29]. **Figure 7** demonstrates small scale applications of anaerobic digestion in households in India.

3.3 Incineration

Incineration is the regulated burning of solid waste with sufficient oxygen under anaerobic conditions at high temperatures above 850°C to release heat. The process also leads to a high-temperature combustion flue gas consisting of CO_2 and water and bottom ash which consists of minimal amounts of leftover carbon [6, 11].

The waste burned can either be in a raw form that is waste immediately after the first three stages of the waste hierarchy or in a pre-treated for like RDF and for each case the plant configuration changes depending on the feedstock. The energy content of raw residue typically ranges from 8 to 11 MJ/kg and the energy content of the pre-treated feed is typically between 12 and 17 MJ/kg [6]. The higher energy content in RDF is because water, recyclable (metals and glass), and inert materials (stones) are removed leaving the waste with the higher energy content [6, 19]. The other advantage of RDF over raw residue is it provides an opportunity to remove most of the hazardous material that could be harmful when burned [19].

The major importance of incineration is to get rid of problematic waste [31]. Incineration decreases the initial quantity and weight of waste by 90% and by 75%, respectively [31] and this makes it suitable for disposing of waste especially in countries

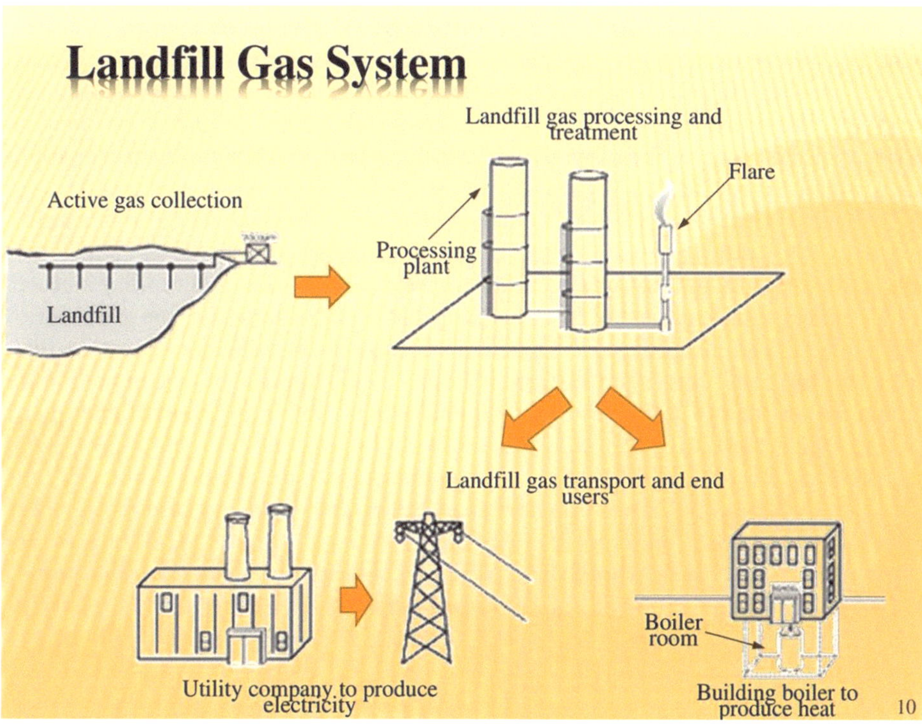

Figure 6.
Gorge Farm Energy Park, Nakuru Kenya [28].

Figure 7.
A family in Maharashtra, India cooking using biogas [30].

that are facing disposal management problems [32]. Typically, 65–80% of the organic waste energy content retrieved as energy. This process uses the combustion heat through a boiler to generate steam. The steam is either applied in steam turbines to produce power and/or used in the heat exchanger technologies to meet heating requirements of an industry or community [6, 19, 24, 33]. A CHP plant that generates heat and/or electricity and is reported to be the most efficient way of recovering energy using a steam boiler [6]. When the incinerator produces heat only, or electricity only, or a combination of both, the efficiency of the plants ranges from 70–80%, 20–25%, and 50–60%, respectively [34]. The choice on whether to produce heat or electricity or both will depend on the needs of the country. The residue bottom ash can be discarded in a landfill or applied as construction material [11]. **Figure 8** demonstrate a CHP incineration plant.

Even though incineration is efficient, the long-term consequences of pollution become evident and suggests the need to improve the fuel compositions, reduce the moisture in the fuel, reduce the sizes of the waste fuel particles, and modify incinerator designs to reduce pollution [24]. The importance of cleaning the flue gas before letting it out in the environment by placing pollution control devices (electrostatic precipitators), or placing an appropriate furnace configuration, or by controlling the combustion process [11]. This flue gas can also be retrieved in the form of energy to generate electricity [24].

Incineration is considered very expensive in terms of capital, and O&M. The process is reported to be more expensive than controlled landfilling and that for the project to economically feasible the energy recovered must be sold. Also, technology is not efficient when the waste composition has low calorific values [33]. **Figure 9** shows a CHP incineration plant in Sweden that handles 700,000 tpa, produces 2174 GWh of heat used in district heating, and 197 GWh of electricity, yearly [36].

3.4 Pyrolysis

Pyrolysis is the thermochemical degradation of organic waste at high temperatures with no oxygen [24, 37, 38]. Also, the process is usually powered by the energy

DOI: http://dx.doi.org/10.5772/intechopen.101904

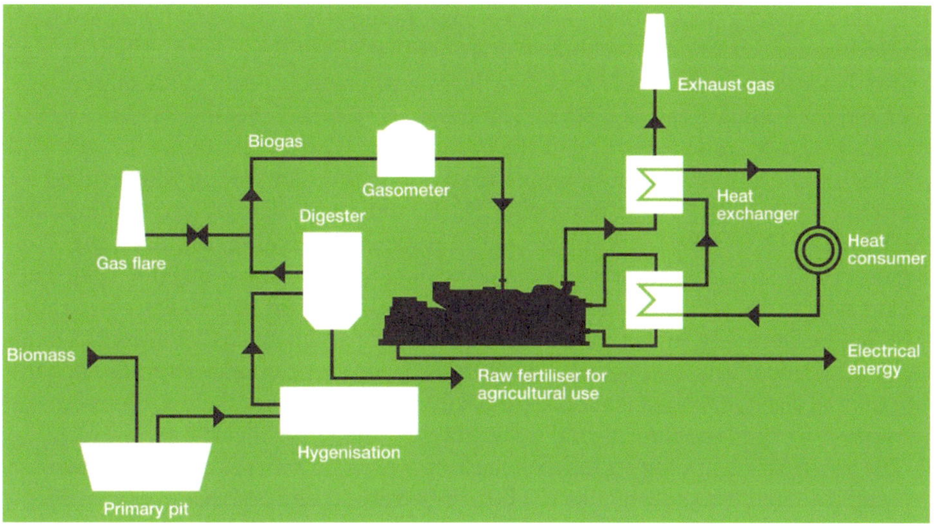

Figure 8.
A schematic illustration of a typical incineration plant [35].

Figure 9.
The Högdalen CHP-plant in Stockholm, Sweden [36].

produced during thermal degradation (endothermic process) [38, 39]. It is reported that the process requires consistent feedstock which limits its commercial-scale applicability from accepting MSW since MSW in its raw form is usually not suitable for pyrolysis and normally would need pre-treatment through mechanical preparation and separation to remove inert materials as well as glass, and metals [6]. However, the process is gaining more attention than incineration because of its ability to use a vast range of industrial and domestic waste and its ability to generate different products

[37]. The by-products are either gases (syngas), liquids (bio-oil), or solids (bio-char) and the process comprises a secondary chamber that where the gases or oils are burned to generate electricity or usable heat [24]. **Figure 10** presents a schematic of pyrolysis plant and the variation in yield depend on parameters such as heating rate, the pyrolytic temperature, and evacuation of the product from the reaction zone [37, 40, 41].

The biochar comprises non-combustible components plus carbon while syngas comprises combustible matter that is CO, H, H_4, and other volatile organic compounds. The bio-oil has high heat value and is used as industrial fuel oil [6, 19]. Also, the products can be a fuel to generate power using gas engines and gas turbines [24]. The CVs for syngas, bio-oil, and biochar range from 10–15 MJ/Nm^3, 15–20 MJ/Nm^3, and 34 MJ/kg, respectively. Even though char's CV is comparable to coal, it is limited by the complex nature of waste which might comprise hazardous elements that pose risk to humans and the environment and care must be taken [37]. Nonetheless, the products are ready to use and specifically, the waste polymers generate the best oil product.

The pyrolysis technology is expensive compared to commercial ways of treatment [24] and the need to pre-treat waste. The pre-treatment devices are expensive and complex [37]. Also, the syngas causes tarring which can easily lead to blockages and operational challenges. Because of this, pyrolysis facilities have been associated with failures and inefficiencies [6]. Failure to sort waste before the process could lead to the production of dangerous nitrogen compounds in the products hence the need gas cleaning devices [37].

Figure 11 shows a pyrolysis plant located in Bulgaria that converts plastic waste into diesel oil [42] and despite the various pilot plants and industrial-scale developments, it is reported [37] the process is still not economically viable.

3.5 Gasification

Gasification is the process through which combustible gas is produced through partially oxidising waste at high temperatures of 800–900°C [24]. Gasification could be considered in-between pyrolysis and combustion because oxygen added neither

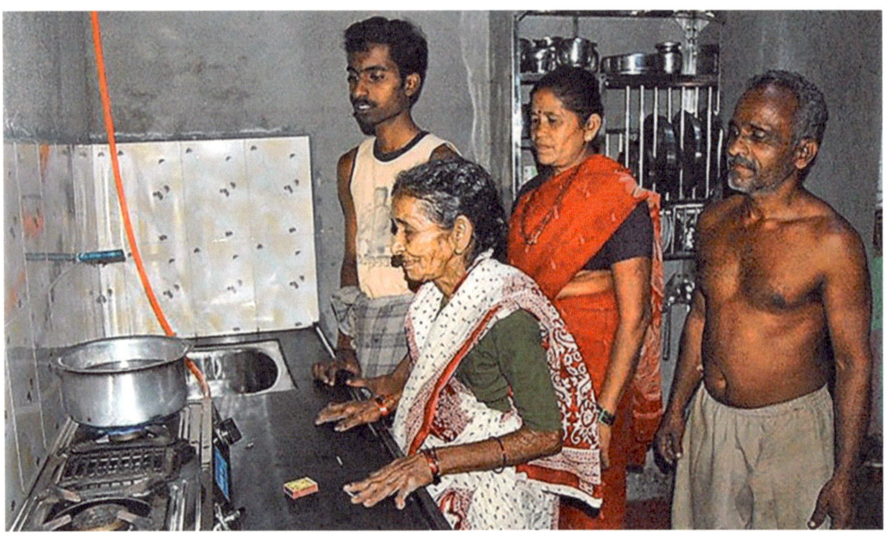

Figure 10.
Illustration of the pyrolysis process [38].

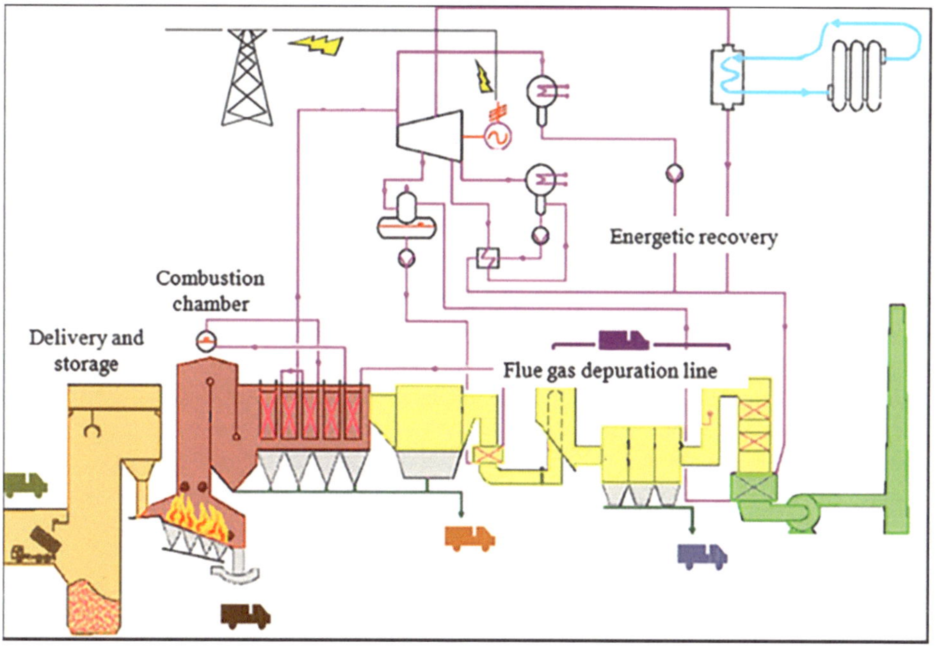

Figure 11.
Huayin plastic to diesel oil plant, Bulgaria [42].

allows full oxidation nor allow full combustion. The gasification process mainly produces its heat however part of the heat is needed to start and continue the process [6]. The gas produced is known as syngas which can be burned to generate heat or in gas engines and gas turbines as a fuel to produce electricity [24]. **Figure 12** illustrates the process of MSW gasification to generate power [43].

The gas generated from gasification is reported to have Net CV varying between 4 and 10 MJ/Nm^3. Another product is a solid residue of ash which is non-combustible and has relatively low levels of carbon [6]. The produced gas can be utilised to generate power using IC engines [19].

The gasification process has advantages of reasonable costs, and flexibility of integrating the working conditions of temperature and equality ratio, and the reactor

Figure 12.
Schematic of MSW gasification to produce power [43].

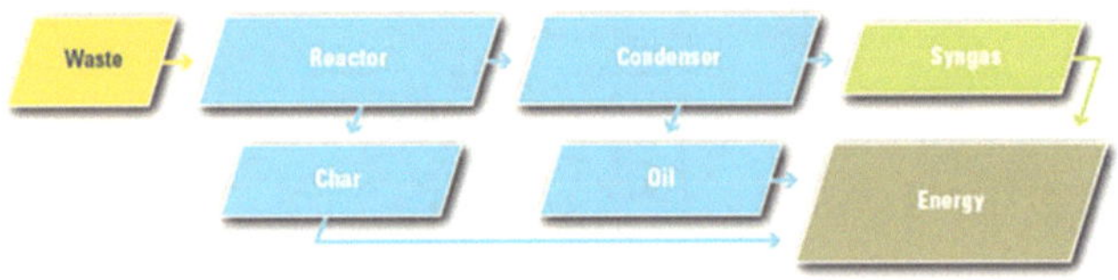

Figure 13.
Waste gasification plant in Lebanon, Tennessee [44].

arrangement to obtain syngas [24]. The process is reported to improve the heating value of gas produced and has lesser quantities of residues compared to incineration and pyrolysis [19].

A report [24] indicates it is associated with the complexity to adapt to various characteristics of different waste and these usually prevent it from commercial applications. The issue is that the source of fuel for gasification will change over time due to variations in waste. Also, MSW in its raw form is usually not suitable for gasification and normally would need pre-treatment through mechanical preparation and separation to remove inert materials as well as glass, and metals [6]. The process suffers disposition of tars that causes blockages leading to operational challenges. This problem has been linked to plants failing and inefficiencies in some pilot and commercial-scale plants. However, it has been observed that when higher temperatures are applied, the tars 'cracks' and generate a relatively clean syngas. The plasma gasification technology is a high-temperature process that is potentially used at different stages in different configurations in the gasification process and the ash generated can be transformed into an inert residue under extremely high-temperature thermal methods. Also, other initiatives are set to ensure that the efficiencies of energy recovery from Gasifiers are maximised by using hydrogen fuel cells and gas engines [6]. **Figure 13** shows a waste gasification plant in Lebanon which has a capacity of 64 tonnes/day to generate 420 kW. The waste (wood and tires) is collected and shredded into 1–3 inches and the sludge is blended on the site as well. The target moisture for post-treated waste is 30% for this specific site [45].

4. Feasibility analysis and the application of the WTE technologies in Uganda

4.1 Uganda's waste generation and management

Uganda's current waste management system involves both the private and public sectors and that the estimated solid waste generation rates for Uganda range between 0.55 and 0.6 kg/person/day [46–48]. The respective generation rates were based on studies done on Kampala district and Mukono district, respectively but this could be comparable to other districts. Further, a study revealed that waste generation rates in Uganda are 0.3 kg/day for low-income homes, and 0.66 kg/day for high-income homes and that the domestic (residential) sector of the country contributes 52% (ref. **Table 1**) of the waste generated [49]. However, this study was carried out from only the 9/15 of urban cities from the political-administrative regions of Uganda.

The different sectors generate mainly organic wastes (food waste) and the dry wastes are the minor forms of waste. Also, the waste composition of the industry sector varies depending on the type and all this data is illustrated in **Table 1** [49]. Results in

Sector	Contribution by weight	Characteristics
Domestic	52	Majority: food wastes
		Minority: plastics, paper, textiles, glass, ceramic, ashes, leather, compound wastes.
Markets	20	Major: vegetable and fruit waste
		Minor: damaged packaging material like sacks and poly-ethene bags.
Commercial minus markets	8	Major: packaging material, food waste, scrap metal
		Minor: glass, hazardous waste, containers
Institutional	5	Major: food waste and stationery
		Minor: packaging material
Industrial	3	Varies depending on the industry
Health care	1	Major: domestic type
		Minor: hazardous wastes
Others	11	Street sweeping, public park and construction waste.

Table 1.
Waste in Uganda.

Table 1 are relative to a study carried out, by Okot-Okumu [46], to assess waste management in three significant towns in Uganda (Kampala, Jinja, and Lira) which revealed that the biodegradable composition of waste was higher that is 77.2, 78.6, and 68.7%, respectively. Also, a study [50] discovered that Kampala generates up to 28,000 tonnes of waste/month with an organic composition of 92.1% while plastic and paper account for only 5.9%. The reason for the difference could be that the former [46] examined solid waste from its origin to final dumping and was carried out using existing publications and reports while the latter [50] was carried out through sampling, field measurement, and laboratory tests of waste disposed at the Kiteezi landfill in Kampala for a year (July 2011–June 2012) to obtain the chemical composition. Nonetheless, both studies could imply that waste in Uganda mainly comprises of organic waste.

The most sought-after way of collecting waste is when waste producers move their waste to community collection sites such as bunkers or skips, and the waste is taken to landfills by the respective municipalities. In some cases, the private sector waste management companies collect waste from house to house but this is normally at a fee, or the larger institutions and commercial businesses hire the private companies to handle the waste [46]. It was found that communities with bad road access are avoided by collecting trucks which leads to high rates of open dumping as a means of disposal by the waste producers [50]. Also, reports point out that apart from poor road access, unaffordability when a waste collection fee is required is another cause of poor solid management [51]. Most of the urban areas in Uganda have waste released in gardens, along the road, open dumps, and channels. **Figure 14** is an illustration of open dumping in urban areas of Uganda [48]. Open dumping poses environmental and health risks for the respective communities through pollution of soil, and surface water, degradation of the ecosystem as well as GHG emission when the organic waste decomposes [51, 53].

Reports [54] reveal that landfilling is the only authorised form of disposal currently in the country and other forms such as open dumping, uncontrolled

Figure 14.
A display of open dumping in Kampala [52].

burning, relative recycling, and composting which take place at unknown extents. Also, 40–45% of waste generated is gathered and thrown away to the landfills and that 11% is recycled by waste pickers [50, 51, 54]. This is comparable to a report that indicates 50% generated waste is collected. For Kampala city, all the waste collected is usually dumped on one landfill, Kiteezi, which is about 12 km from the city centre [50, 55]. This landfill is a sanitary landfill occupying 0.146 km^2 of land with a leachate treatment system that decreases the biological oxygen before the leachate is discarded to the nearby wetland. Despite this initiative, residents around complain of bad odour, leachate leakages, and increased scattering of wastes by marabou stocks causing their land to lose value. Also, the openness and mismanagement of the landfills cause a problem of air pollution through GHG emissions which pose a health risk [46, 50].

The 2010 audit report for SWM in Kampala credits the inefficiencies in waste management to the lack of awareness which has led to aimless littering and uncontrolled burning of waste [56]. A study [46] points out that the record of waste collected accounts for that which reaches the community collection points and the uncollected waste is not recorded. This could be problem when identifying suitable management techniques due to insufficient waste data.

The mentioned studies are specific to just a few cities in the country with a major focus on the capital city Kampala. However, it is plausible to conclude that majority of waste produced in Uganda comprises organic waste with an overall composition of above 70%. This could also be assumed since the major economic activity is agriculture which is usually associated with organic waste.

A study [46] reports that the majority of the waste is mixed and there is no official structure of sorting waste in the country. Sorting may happen when workers segregate wastes of value on the way to the landfills or at the waste bunkers, road verges, skips, or at the landfills and this is illustrated in **Figure 15**. Most wastes hand-picked are plastics comprising of jerry cans, and bottles as well as cardboard. In some cases, the separation is done only when the producer is looking to reuse the plastic material or glass bottles or use food leftovers as animal feeds [46, 51]. **Figure 16** is an image of plastic bottles collected for recycling.

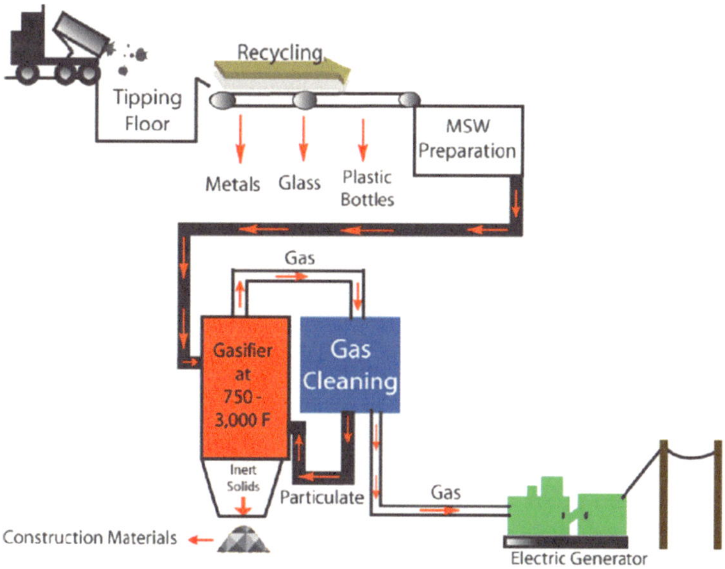

Figure 15.
Waste picker segregating waste at Kiteezi landfill [57].

Figure 16.
Coca-Cola recycling plant at Kyambogo-Kampala Uganda [55].

4.2 Technology capabilities in Uganda

The choice on which technology to adapt depends on the local conditions and energy requirements of the communities and/or sectors of a country. Because of these reasons, the Government of the UK always maintains an attitude of being technology-neutral when promoting private investment unless the technology shows evidence of market failure [7]. Also, knowledge of the organic fraction, calorific value, and chemical composition enables a country to know how best to manage the waste [11].

Therefore, the successful implementation of anaerobic digestion in Uganda would largely depend on waste generated from agriculture and reports that the residential and market sectors generate more than 72% of the waste which largely comprises of food waste. The waste from these sectors has a high moisture content which would accelerate production of biogas.

It would be suitable to meet the energy needs of these market structures while solving the problem of food waste management. Also, in Section 3 it is mentioned that the yield of anaerobic digestion is higher when food waste is fed into the digesters along with MSW could make this technology a reliable source of energy in residential households, and markets in Uganda since they generate mostly food waste. The country could adopt large scale anaerobic digestion like the Gorge anaerobic digestion plant in Kenya highlighted in (Section 3) where the waste from farms is used to generate electricity for farm activities. In the long run, the biogas generated from large scale projects can used in the transportation sector as a source of fuel which could introduce flexi-fuel vehicles that use both petroleum and bio-methane [58].

Regarding WTE incineration, the technology is more efficient when the CV of the waste is high. Uganda's waste has a CV of 6.2 MJ/kg which is below the typical range for raw waste highlighted in Section 3 hence the country could apply the circulating fluidised bed combustion technology which permits waste with low CV. However, this type of incinerator processes lower quantities of waste compared to the grate-based combustion technology. An alternative would be to pre-treat the waste to increase the efficiency of the plant. Overall, the application of incineration would generate electricity that would meet the demands of manufacturing industries and surrounding communities to promote energy security. For example, a similar project like the Reppie plant in Ethiopia could be set up in Uganda to process waste to generate about 20 MW of electricity. Such a project is comparable because Ethiopia is reported to have waste compositions comprising of 60% organic waste [59, 60] which is similar Uganda. Also, this plant has a pre-treatment section to increase the energy content in the waste. Such a plant could process waste from Kampala which is estimated to generate 28,000 tonnes/month ≈ 930 tonnes/day [50]. Such projects could be implemented around the country to reduce the quantity of waste that goes to the landfills and to improve energy security.

In Section 3 it is noted that the incineration plant is more efficient when it generates heat or both heat and power. The heat produced may be wasted since district supply heating systems are not necessary as the country's temperatures are relatively warm throughout the year (26°C [61]) hence household or commercial heating is not required. However, the heat produced can be used for heating processes in nearby factories.

Concerning landfill gas recovery, this technology would require sufficient land to implement. In Section 3 an average landfill site occupies 600 acres but with the current size of Uganda and the high rate of population increase, the application of landfill gas recovery would be affected by shortage of land. Also, the landfill sites would have to be away from the growing cities due to the high rates of urbanisation in the country. Nonetheless, this mechanism would still be applicable in parts of the country that are less populated and have sufficient land. This would however incur costs to transport waste generated to such locations. Comparing landfills and anaerobic digestion, the former generates lower fractions of methane than the latter and so it would be advisable to consider anaerobic digestion.

Turning now to pyrolysis and gasification, Section 3 reveals that the processes are difficult to scale up hence cannot be used for large scale purposes and commercial purposes due to their complexities in fuel requirements. However, these technologies could be adopted by industries, specifically the rotary kiln type of reactor which offers the advantage proper heat transfer allowing it to process waste polymers (plastics) which have low thermal conductivity and generate the best quality oil suitable for industrial purposes. However, Uganda has low generation rates of waste polymers (dry waste) and projections indicate low dry waste generation in the next decade hence such technologies could be unsustainable in the long run.

Regarding waste as a fuel, [7] the waste needs to be appropriate for the technology in question. Reports state that the inadequacy in the supply of feedstock to biogas plants is a barrier to the technologies. This could be comparable to the other technologies [62]. Anaerobic digestion, incineration, and pyrolysis need the waste to be pre-treated through separation, sorting, processing and mixing with additives to optimise efficiency, increase the calorific values, and reduce levels of pollution [31]. In observed that fossil fuel-based wastes emit GHG which contribute to climate change, which is another reason to why waste should be separated. Literature reports that lack of waste separation has led to the failure of large-scale bio-methanation in India [63]. Therefore, Uganda would have inconsistencies in the waste fuel due to lack of separation leading to mixed wastes and this could affect the sustainability in the long run. This will also increase costs incurred by the plants to mechanically treat waste. Nonetheless, these technologies could still be adopted by manufacturing industries in Uganda which are more consistent in the characteristics of waste as per industry, for example, Kakira Sugar factory processes bagasse to produce heat and electricity.

Also, lack of separation could lead to hazardous materials in the waste which pose the risk of generating toxic chemicals in solid and gaseous residues from the processes that are later disposed of. However, Uganda has low levels of industrialisation and it would seem unlikely to have large quantities of harmful material in the waste which could make mechanical separation easier. In addition, Uganda would need to improve her collection efficiencies from 45% to the daily target of 80% to ensure the plants have a consistent supply of fuel. A study [24] claims that when technology can work with inconsistent fuels then it is feasible for such communities and in this case landfill gas recovery would be the most suitable.

Lastly it is noted in Section 3 that the different designs and configurations of anaerobic digestion require the use of water to optimise the digestion of MSW. This is a could be a major problem since reports [53] show that the availability of water is limited in densely populated regions of Uganda and that 76% of Ugandans have water within reach of 1 km. To solve this, the country could focus on using the high solid continuous digestion systems which require little water (Section 3(1)). The supply of biogas is inadequate to meet the needs of a community, they resort to the rudimentary sources of fuel [62]. This concern would be comparable to the other WTE technologies as the choice by beneficiaries to adopt any technology is most likely dependent on reliability.

4.3 Policies review

Regarding consistency, that waste-sorting has a significant effect on the efficiency of all technologies and since Uganda lacks waste separation regulations this will affect the output of the technologies which will in turn affect the revenue flows [64]. The National Environment (Waste Management) Regulations, S.I. No 52/1999 under sections 53(2) and 107 of the National Environment Act, Cap 153 [65], shows no mandate which directs a waste producer to explicitly separate the waste generated according to physical or chemical composition. This causes a problem of delivering mixed waste that makes it hard for plants to have consistency in the physical and chemical composition of waste which could affect the sustainability of plants in the long run. A lot of effort would then be needed to sort waste before extracting the energy and these extra costs may not be attractive to investors. Also, these extra costs may lead to a rise in the cost of electricity purchased by the customers. To mitigate these, the GOU could enforce some of the UK's policies (**Table 2**) to enhance better SWM. Also, awareness campaigns through media platforms and community focus groups can help solve the problem of waste segregation to improve the efficiency of WTE initiatives [62].

	Policy	Impact
1	Keep waste at minimum through prevention, recycling, or recovery through policies like placing charges on carrier bags has pushed for recycling within communities	This could prevent WTE from competing with recycling
		It could lessen the volume of waste sent to landfills
2	Store waste securely, use suitable containers to avoid leakages, ensure containers are waterproof. Containers must be labelled clearly to indicate the type of waste contained.	This will enhance waste separation allowing WTE technologies to have consistent fuels
		Promotes consistency in waste fuels and eases the mechanical separation of the WTE plants
3	Store different types of waste separately to avoid contamination, and to permit reuse and recovery	This will enhance waste separation allowing WTE technologies to have consistent fuels
4	Waste producer must classify their waste to the waste contractor before sending it for disposal or recycling	Promotes consistency in the waste fuels and eases the mechanical separation at WTE plants
5	It is illicit to mix hazardous and non-hazardous waste and there are guidelines to enable a waste producer to identify the types.	Mitigates the pollution that would arise
		Protects the environment from harmful substances
6	The Government of the UK provides a platform to report fly-tipping (illegal waste dumping). In Northern Ireland, it is required to report a waste producer who intentionally labels waste inappropriately. Also, the law permits reporting littering along local streets	Reduce open dumping
		Ensures that all waste is collected appropriately allowing plants to have enough waste fuel
7	For an operator to carry out waste treatment, some rules and regulations that must be followed to protect the environment	Ensures all waste is collected appropriately allowing plants to have enough waste fuel
		Ensures pollution control systems are in place

Table 2.
A review of some of the UK's waste management regulations [66].

5. Recommendations for further research

A limitation of this study is the lack of sufficient data on SWM in Uganda. The research has identified that little effort has been directed towards the implementation of WTE in Uganda as most studies have been focused on the waste generation and composition and specific to just a few major cities in the country. Another limitation is that majority of the data is not up to date. Greater efforts are needed to ensure that more resources are directed towards SWM studies all over the country with a special emphasis on promoting designs and programs to mitigate the weaknesses in the system as a way of easing the work of potential developers.

An alternative area of research could be a case when the GOU partly or fully finances the projects to waive the financial burden on the citizens. With this, the government could register the projects under CDM and gain carbon credits that can be sold to the open market. Also, such projects would be associated with job creation, improved energy security, and reduced deforestation. It is important to realise that the economics and environmental benefits could be difficult to evaluate by the GOU however success in SWM by a municipality was noted as a guarantee to thrive in other sectors.

6. Conclusion

The purpose of this chapter was to evaluate how and why Uganda could adopt WTE technologies to reduce waste volumes. The continual increase in the waste generation rate is evidence that there is a need for Uganda to assess different waste management techniques. The proposed WTE techniques are landfill gas recovery, anaerobic digestion, incineration, pyrolysis, and gasification, and the following are the observations.

Firstly, all WTE technologies are applicable in Uganda because the different sectors differ in energy needs. From this, the GOU needs to remain technologically neutral when promoting WTE. Anaerobic digestion would be the most reliable since most of the waste in Uganda is organic waste mainly from agricultural activities. It is also feasible in residential homes on a small scale to meet heating and cooking needs and in turn, would reduce the use of wood fuels which are associated with health risks and deforestation. Incineration is also considered but because of the reported low CV of Uganda's waste, such a plant would need to pre-treat the waste to increase the CV. Gasification and pyrolysis are noted to be more advanced and better in terms of the products but difficult to scale up due to the technological limitations and their complexity in adapting to inconsistent waste. However, pyrolysis and gasification could be applicable in industries to improve energy security and reduce the demand on the national grid. Landfill gas recovery could be applied at any abandoned site, but it would be limited by a shortage of land near the cities especially with the growing populations. Also, the by-products like biogas and bio-oil could be adopted in the transportation sector in the long run. This would reduce the country's dependency on imported fuels.

Secondly, it is noted that the sustainability of these technologies is greatly affected by the composition of waste as a fuel and the waste must be consistent in physical and chemical composition. Consistency can be guaranteed when Uganda improves waste collection techniques by enforcing the proposed laws and regulations that promote waste separation and efficient waste collection. This would also ensure that WTE plants have enough waste to maximise efficiency. Also, the separation of waste is key in limiting pollution associated with the thermo-chemical processes since it enables the removal of hazardous material.

Finally, WTE must not rival prevention, recycling, and reuse but should complement them when the possibilities are exhausted. This could in turn decrease the amount of waste taken to the landfills. Also, WTE initiatives should not be used as an excuse to generate waste. The country could adopt the waste management hierarchy through awareness programs and enforcing laws and regulations to impact people's behaviours and attitudes leading to reduced quantities of waste generated, to mitigate the issue of open dumping and other associated issues rising from poor waste management.

Author details

Charlene Nagawa[1,2]

1 Sustainable Energy and Entrepreneurship, University of Nottingham, Nottingham, UK

2 Equatorial Power, Kampala, Uganda

*Address all correspondence to: charlenenagawa@gmail.com

References

[1] Ellis C. World Bank: Global Waste Generation Could Increase 70% by 2050 [Online]. 2018. Available from: https://www.wastedive.com/news/world-bank-global-waste-generation-2050/533031/ [Accessed: 24 August 2020]

[2] NASA. Global Climate Change [Online]. 2020. Available from: https://climate.nasa.gov/ [Accessed: 30 July 2020]

[3] Ritchie H. Food Waste is Responsible for 6% of Global Greenhouse Gas Emissions [Online]. University of Oxford; 18 March 2020. Available from: https://ourworldindata.org/food-waste-emissions [Accessed: 15 June 2020]

[4] Thorium Energy World. World Energy Situation [Online]. IThEO; 2018. Available from: http://www.thoriumenergyworld.com/energy.html# [Accessed: 15 June 2020]

[5] BBC. Reasons for Increase in Demand for Energy [Online]. BBC; 2020. Available from: https://www.bbc.co.uk/bitesize/guides/zpmmmp3/revision/2 [Accessed: 16 June 2020]

[6] DEFRA.UK. Incineration of Municipal Solid Waste [Online]. GOV.UK; 27 February 2013. Available from: https://assets.publishing.service.gov.uk/government/uploads/system/uploads/attachment_data/file/221036/pb13889-incineration-municipal-waste.pdf [Accessed: 23 June 2020]

[7] DEFRA.UK. Energy from Waste: A Guide to the Debate [Online]. February 2014. Available from: https://assets.publishing.service.gov.uk/government/uploads/system/uploads/attachment_data/file/284612/pb14130-energy-waste-201402.pdf [Accessed: 21 July 2020]

[8] Maplecroft V. Waste Generation and Recycling Indices 2019: Overview and Findings [Online]. June 2019. Available from: https://www.circularonline.co.uk/wp-content/uploads/2019/07/Verisk_Maplecroft_Waste_Generation_Index_Overview_2019.pdf [Accessed: 14 June 2020]

[9] The Africa Report. Now is the Time to Turn Africa's Waste into Energy [Online]. 2019. Available from: https://www.theafricareport.com/18961/now-is-the-time-to-turn-africas-waste-into-energy/ [Accessed: 2020 August 2020]

[10] DEFRA.UK. Guidance on Applying the Waste Hierarchy [Online]. June 2011. Available from: https://assets.publishing.service.gov.uk/government/uploads/system/uploads/attachment_data/file/69403/pb13530-waste-hierarchy-guidance.pdf [Accessed: 18 June 2020]

[11] Awasthi MK, Zhao J, Soundari PG, Kumar S, Chen H, Awasthi SK, et al. Chapter 6: Sustainable management of solid waste. In: Sustainable Resource Recovery and Zero Waste Approaches. Elsevier B.V; 2019. pp. 79-99. Available from: https://www.elsevier.com/books/sustainable-resource-recovery-and-zero-waste-approaches/taherzadeh/978-0-444-64200-4

[12] Hoornweg D, Bhada-Tata P. What a Waste: A Global Review of Solid Waste Management. Washington, DC: The World Bank; 2012

[13] Idris A, Inanc B, Hassan MN. Overview of waste disposal and landfills/dumps in Asian countries. Journal of Material Cycles and Waste Management. 2004;**6**:104-110. DOI: 10.1007/s10163-004-0117-y

[14] Ritchie H, Roser M. Energy [Online]. University of Oxford; 2015. Available from: https://ourworldindata.org/energy#citation [Accessed: 16 June 2020]

[15] Chu L. Landfills. In: Encyclopedia of Ecology. Elsevier B.V; 2008. pp. 2099-2103. Available from: https://www.sciencedirect.com/topics/earth-and-planetary-sciences/landfill

[16] Derbyshire.Gov.UK. How Landfill Sites Work [Online]. Derbyshire County Council; 2020. Available from: https://www.derbyshire.gov.uk/environment/rubbish-waste/waste-strategy/landfill/how-landfill-sites-work/how-landfill-sites-work.aspx [Accessed: 20 June 2020]

[17] EastSussex.Gov.UK. What is a Landfill Site [Online]. East Sussex County Council; 2020. Available from: https://www.eastsussex.gov.uk/environment/rubbishandrecycling/landfill/whatarethey/ [Accessed: 20 June 2020]

[18] Corbley KP. Mapping Landfill Space NASA Style [Online]. 2001. Available from: https://www.waste360.com/mag/waste_mapping_landfill_space [Accessed: 31 August 2020]

[19] CPHEEO. Chapter 15: Energy recovery from municipal solid waste. In: Manual on Municipal Solid Waste Management—2000. Ministry of Housing and Urban Affairs, Government of India; 2019. p. 48. Available from: http://cpheeo.gov.in/upload/uploadfiles/files/Part2.pdf

[20] Vaverková MD. Landfill Impacts on the Environment—Review [Online]. 3 October 2019. Available from: https://www.mdpi.com/2076-3263/9/10/431/htm [Accessed: 21 July 2020]

[21] SlidePlayer. Landfill Gas Collection and Recovery [Online]. 2020. Available from: https://slideplayer.com/slide/12325423/ [Accessed: July 2020]

[22] Dando T. Viridor Invests £2 Million to Transform Landfill Gas into Transport Fuel [Online]. 2019. Available from: https://resource.co/article/viridor-invests-2-million-transform-landfill-gas-transport-fuel [Accessed: July 2020]

[23] Viridor. Landfill Disposal and Landfill Gas Recovery [Online]. 2020. Available from: https://www.viridor.co.uk/energy/landfill-disposal-and-landfill-gas-recovery/ [Accessed: 29 June 2020]

[24] Rajendran K, Lin R, Wall DM, Murphy JD. Chapter 5: Influential aspects in waste management practices. In: Sustainable Resource Recovery and Zero Waste Approaches. Elsevier B.V; 2019. pp. 65-78. Available from: https://www.elsevier.com/books/sustainable-resource-recovery-and-zero-waste-approaches/taherzadeh/978-0-444-64200-4

[25] Benato A, Macor A, Rossetti A. Biogas engine emissions: Standards and on-site measurements. Energy Procedia. 2017;**126**:398-405

[26] ClarkeEnergy. Biogas [Online]. 2020. Available from: https://www.clarke-energy.com/applications/biogas/ [Accessed: July 2020]

[27] Nixon J, Dey P, Ghosh S, Davies P. Evaluation of options for energy recovery from municipal solid waste in India using the hierarchical analytical network process. Energy. 2013;**59**:215-223

[28] Kiganda A. Africa's First Grid Connected Biogas Farm in Kenya Powers Up [Online]. 2017. Available from: https://constructionreviewonline.com/2017/01/africas-first-grid-connected-biogas-farm-in-kenya-powers-up/ [Accessed: July 2020]

[29] Kamadi G. Africa's First Grid-Connected Biogas Plant Powers Up

[Online]. 2017. Available from: https://www.reuters.com/article/kenya-energy-biogas/africas-first-grid-connected-biogas-plant-powers-up-idUSL5N1EZ1KL [Accessed: 02 August 2020]

[30] Bose HK. Multimedia: Biogas Eases Pressure on Maharashtra Forests [Online]. 2019. Available from: https://indiaclimatedialogue.net/2019/02/25/biogas-eases-pressure-on-maharashtra-forests-a-multimedia-report/ [Accessed: July 2020]

[31] Tabasová A, Kropác J, Kermes V, Nemet A, Stehlík P. Waste-to-energy technologies: Impact on environment. Energy. 2012;**44**:146-155

[32] What is Incineration? Busch Systems [Online]. Available from: https://www.buschsystems.com/resource-center/knowledgeBase/glossary/what-is-incineration [Accessed: 20 June 2020]

[33] The International Bank for Reconstruction and Development/ The World Bank. Decision Makers' Guide to Municipal Solid Waste Incineration [Online]. 1999. Available from: https://web.mit.edu/urbanupgrading/urbanenvironment/resources/references/pdfs/DecisionMakers.pdf [Accessed: 22 June 2020]

[34] Planete Energies. Total Energies Foundation. Incineration—The Heating Power of Refuse [Online]. 2015. Available from: https://www.planete-energies.com/en/medias/close/incineration-heating-power-refuse#:~:text=This%20process%20is%20energy%20efficient,energy%20per%20ton%20of%20refuse.&text=The%20energy%20efficiency%20of%20this,(300%20to%20400%20kWh) [Accessed: 20 July 2020]

[35] Panepinto D, Zanetti M. Municipal solid waste incineration plant: A multi-step approach to the evaluation of an energy-recovery configuration. Waste Management. 2017;**73**:332-341

[36] Fortum. Högdalen CHP Plant [Online]. 2020. Available from: https://www.fortum.com/about-us/our-company/our-energy-production/our-power-plants/hogdalen-chp-plant#:~:text=In%20Sweden%2C%20the%20H%C3%B6gdalen%20CHP,is%20provided%20by%20Stockholm%20Exergi [Accessed: 02 August 2020]

[37] Czajczynska D, Anguilano L, Ghazal H, Krzyzynska R, Reynolds A, Spencer N, et al. Potential of pyrolysis processes in the waste management sector. Thermal Science and Engineering Progress. 2017;**3**:171-197

[38] Klean Industries. Pyrolysis [Online]. 2020. Available from: http://www.kleanindustries.com/s/Pyrolysis.asp [Accessed: 02 August 2020]

[39] Twigg R, Cressell L, Buchdahl J. Waste [Online]. 2002. Available from: https://www.lordgrey.org.uk/~f014/usefulresources/aric/Resources/Fact_Sheets/Key_Stage_4/Waste/pdf/Waste.pdf [Accessed: 25 July 2020]

[40] Basu P. Biomass Gasification, Pyrolysis and Torrefaction: Practical Design and Theory. 3rd ed. Academic Press; 2018. Available from: https://www.elsevier.com/books/biomass-gasification-pyrolysis-and-torrefaction/basu/978-0-12-812992-0

[41] Zhang Y, Cui Y, Chen P, Liu S, Zhou N, Ding K, et al. Chapter 14—Gasification technologies and their energy potentials. In: Sustainable Resource Recovery and Zero Waste Approaches. Elsevier B.V; 2019. pp. 193-206. Available from: https://www.elsevier.com/books/sustainable-resource-recovery-and-zero-waste-approaches/taherzadeh/978-0-444-64200-4

[42] Huayin. Waste Plastic to Diesel Oil Plant in Bulgaria [Online]. 2019. Available from: http://www.huayinrecycling.com/index.php/home/index/case_detail/id/26 [Accessed: 02 August 2020]

[43] Zafar S. Gasification of Municipal Solid Waste [Online]. January 2009. Available from: https://www.researchgate.net/publication/285295261_Gasification_of_municipal_solid_wastes [Accessed: 02 August 2020]

[44] E Leader. Lebanon's Gasification Plant is Blending Waste and Generating Sustainable Power [Online]. 2020. Available from: https://www.environmentalleader.com/projects_of_year/aries-clean-energy/ [Accessed: July 2020]

[45] E+E Leader. Aries Clean Energy [Online]. 2020. Available from: https://www.environmentalleader.com/projects_of_year/aries-clean-energy/ [Accessed: 02 August 2020]

[46] Okot-Okumu J. Solid Waste Management in African Cities—East Africa [Online]. October 2012. Available from: https://www.intechopen.com/books/waste-management-an-integrated-vision/solid-waste-management-in-african-cities-east-africa [Accessed: 07 July 2020]

[47] KCCA [Online]. October 2017. Available from: https://www.kcca.go.ug/uDocs/kampala-waste-treatment-and-disposal-ppp.pdf [Accessed: 27 July 2020]

[48] Yusufa AA, Onu P, Hassan AS, Tunji LA, Oyagbola IA, Mustafa MM, et al. Municipality solid waste management system for Mukono District, Uganda. Procedia Manufacturing. 2019;**35**:613-622

[49] Okot-Okumu J, Nyenje R. Municipal solid waste management under decentralisation in Uganda. Habitat International. 2011;**35**:537-543

[50] Komakech AJ, Banadda NE, Kinobe JR, Kasisira L, Sundberg C, Gebresenbet G, et al. Characterization of municipal waste in Kampala, Uganda. Journal of the Air & Waste Management Association. 2014;**64**(3):340-348. DOI: 10.1080/10962247.2013.861373

[51] Mukama T, Rawlance N, Musoke D, Musinguzi G, Halage AA, Carpenter DO, et al. Practices, concerns, and willingness to participate in solid waste management in two urban slums in Central Uganda. Journal of Environmental and Public Health. 2016;**2016**:7. DOI: 10.1155/2016/6830163

[52] Balcony Times. Community Service, the Impeccable Answer to Disentangle the Garbage Problem in Developing Towns [Online]. 2017. Available from: http://thebalconytimes.blogspot.com/2017/01/community-service-impeccable-answer-to.html [Accessed: July 2020]

[53] Smith DJ. The Potential of Small-Scale Biogas Digesters to Alleviate Poverty and Improve Long Term Sustainability of Ecosystem Services in Sub-Saharan Africa [Online]. 2011. Available from: https://assets.publishing.service.gov.uk/media/57a08ad9e5274a31e00007ec/FinalReport_Biogas-Digesters-in-Sub-Saharan-Africa.pdf [Accessed: 21 July 2020]

[54] Oyoo R, Leemans R, Mol APJ. The determination of an optimal waste management scenario for Kampala, Uganda. Waste Management & Research. 2013;**31**(12):1203-1216. DOI: 10.1177/0734242X13507307

[55] AMREF. Turning the Tide on Plastic "Here Just in Time" The Power of Partnerships [Online]. 2019. Available

from: https://amref.org/uganda/news/turning-tide-plastic-just-time-power-partnerships/ [Accessed: 04 August 2020]

[56] Republic of Uganda. Value for Money Audit Report on Solid Waste Management in Kampala City Council. Kampala, Uganda: Office of the Auditor General; 2010

[57] Schoebitz L. Kiteezi Fall [Online]. 2014. Available from: https://www.flickr.com/photos/110829077@N08/15518195699 [Accessed: July 2020]

[58] Sawin J. ScienceDirect Flex-Fuel Vehicle [Online]. 2012. Available from: https://www.sciencedirect.com/topics/engineering/flex-fuel-vehicle [Accessed: 30 March 2020]

[59] Ethiopia. Overview of Addis Ababa City Solid Waste Management System [Online]. February 2010. Available from: https://www.un.org/esa/dsd/susdevtopics/sdt_pdfs/meetings2010/icm0310/2b-2_Tessema.pdf [Accessed: 15 July 2020]

[60] Assefa M, Mohammed M. Solid waste generation rate and characterization study for Laga Tafo Laga Dadi Town, Oromia, Ethiopia. International Journal of Environmental Protection and Policy. 2017;**6**(2330-7536):84-93

[61] Weather & Climate. Climate and Average Weather in Uganda [Online]. 2020. Available from: https://weather-and-climate.com/average-monthly-Rainfall-Temperature-Sunshine-in-Uganda#:~:text=Uganda%20is%20sunny%20most%20of,May%20and%20October%20till%20November [Accessed: July 2020]

[62] Mittala S, Ahlgrena EO, Shuklab P. Barriers to biogas dissemination in India: A review. Energy Policy. 2018;**112**:361-370. DOI: 10.1016/j.enpol.2017.10.027

[63] Kalyani KA, Pandey K. Waste to energy status in India: A short review. Renewable and Sustainable Energy Reviews. 2014;**31**:113-120

[64] Lim WB, Yuen E, Bhaskar AR. Waste-to-Energy: Green Solutions for Emerging Markets [Online]. 2020. Available from: https://home.kpmg/xx/en/home/insights/2019/10/waste-to-energy-green-solutions-for-emerging-markets.html [Accessed: 29 July 2020]

[65] KCCA. The National Environment (Waste Management) Regulations [Online]. 19 October 1999. Available from: https://www.kcca.go.ug/uDocs/waste_management_regulations.pdf [Accessed: 28 July 2020]

[66] GOV.UK. Environmental Management: Waste [Online]. 2020. Available from: https://www.gov.uk/topic/environmental-management/waste [Accessed: 28 July 2020]

Section 4

Engineering Practices

Chapter 8

Engineering Measures for Isolation and Sequestration of Heavy Metals in Waste as Safe Final Sink

Tomonori Ishigaki, Hiroyuki Ishimori, Hiroki Kitamura and Masato Yamada

Abstract

The long-term safety management of hazardous substances is essential to the development of an environmentally sound resource circulation society. To achieve this, engineering measures to attenuate environmental risks in the isolation and sequestration of hazardous heavy metals are reviewed. From the standpoint of the isolation and sequestration of heavy metals from resource circulation, we assess the challenges in implementing immobilization technologies, constructing updated isolation structures, and controlling environmental conditions. It is also focused intensively on the (bio) chemical transformation behavior of heavy metals and its effect on the migration of the transformed materials in the environment. The contributions of solubilized and gasified metal components to emission into the environment are considered. The obtained results underscore the necessity of multiple barriers to retard and attenuate the migration of hazardous heavy metals. Innovative schemes for the isolation and sequestration of heavy metals will lead to higher levels of safety and environmentally sound resource circulation.

Keywords: safe resource circulation, multi-barrier approach, final sink, heavy metals, immobilization, retarding migration, adsorption

1. Introduction

The world is facing several global and local problems caused by imbalanced resource utilization and inappropriate handling of waste. Climate change is threatening the lives of vulnerable people and regions by increasing extreme weather events. Marine plastic litter is widely recognized as a major risk to maritime activity, fisheries, and wildlife. Waste management practices, which are regarded as one of the major causes of these problems, must be updated to improve the situation.

Achieving environmentally sound resource circulation is a possible solution. The core concept of environmentally sound resource circulation is the harmonization of industrial (human) society with the earth's natural circulation system. It includes measures, such as (a) increased resource circulation to slow down and ultimately close the resource loop by reducing new resource inputs, (b) ethical production of goods and services that generate hardly any waste to narrow and dematerialize the resource loop, and (c) removal of hazardous substances from the resource circulation loop for the safe loop of resources. Toxic-free resource circulation is essential for the sustainable operation of environmentally sound resource circulation. While it will be achieved, the amount of waste/residue generated can be minimized, and hazards will be removed from the loop of circulation. This minimized residue containing concentrated hazardous substances must be adequately managed to end the negative aspects of these resources.

This chapter describes an engineering approach to isolate and sequester hazardous substances concentrated in waste to remove them from the circulation loop, which will be essential to achieve environmentally sound resource circulation.

2. Concept of sequestration/isolation in environmentally sound resource circulation

The main objectives of the sequestration/isolation of hazardous substances are to reduce and avoid the long-term exposure of humans and the environment to hazardous substances. Environmentally sound resource circulation entails challenges for the actors involved, such as requirements that manufacturers implement cleaner and resource-efficient production. It also means customers should purchase long-life products and recycle their waste. Recycling activity should include only the safe circulation of resources, and hazardous substances must be sequestrated from human activity.

The scheme of sequestration of hazardous substances should be consistent with the strategy for a final sink [1]. A final sink is a process in a manner that satisfies the acceptance level of a substance flow as low as the natural environment and the acceptance level of exposure of substances for human health. Hazardous organic chemicals are not expected to be mineralized, and it is hard to secure their long-term immobilization in an isolation site (or a landfill to be upgraded for sequestration). Therefore, the final sink for these substances must occur through physical or chemical destruction. This indicates that the potential of an isolation site to be a final sink is limited to toxic heavy metals or inorganic substances.

Isolation/sequestration should be designed under the multi-barrier approach. The major elements of this approach are (i) containment of toxic substances by stabilization and insolubilization with chemical or physical measures, (ii) avoidance of the release of toxic substances from containment, including exposure to moisture, (iii) retardation of the migration of hazardous substances within a site and the environment, and (iv) early warning of the potential release of a toxic substance by monitoring of containment structures, gases/leachates, and the environment.

3. Containment of toxic heavy metals for safe sequestration

Most countries have legal regulations for the disposal of waste containing hazardous heavy metals. Waste that meets certain criteria can be disposed of in landfills equipped with emission control measures (e.g., leachate treatments and gas collection) [2, 3]. Otherwise, those waste products must be delivered to facilities with containment functions. Pretreatment to detoxify and immobilize waste containing heavy metals is an essential measure to reduce potential emissions reasonably.

Air pollution control (APC) residues, which are generated through thermal treatments of waste, such as incineration, gasification, and pyrolysis, are commonly classified as hazardous materials owing to the high leaching potential of toxic metals. Before APC residues are disposed of, pretreatments are required in many countries to prevent the release of toxic metals into the environment [4]. The available pretreatments can be categorized into three groups—(i) physical or chemical separation, (ii) solidification/stabilization (S/S), and (iii) thermal treatment [5]. Chemical stabilization using organic chelating agents, such as piperazine-based or dithiocarbamate-based agents, is often preferred because such treatments are simple, do not require pretreatments, such as pH control, and remain stable across a wide pH range [6, 7]. On the other hand, these treatments are significantly more expensive than other forms of chemical stabilization [8]. In addition, organic components derived from chelating agents induce long-term leachate problems treatments at landfill sites [9].

3.1 Cement solidification

Cement solidification is a widely used containment technique around the world. The purpose is to avoid the leakage of toxic substances into the environment. The target chemical substances are diverse, such as heavy metals, F, B, and even radioactive substances. Hiraoka and Takeda [10] investigated the effects of cement solidification on the leaching amounts of Hg and Cd in waste sludges in relation to compressive strength. They suggested that the solidification of landfill waste containing heavy metals is safe when the cement amount is over 150 kg/m^3 and the compressive strength is over 0.98 MPa.

Cement solidification was also applied to radioactive cesium-contaminated APC residue generated after the Fukushima Daiichi nuclear disaster in 2011. Radioactive cesium is hardly precipitated in the alkaline condition in a cement mixture and cannot be chemically stabilized, although the solubilities of heavy metals are reduced in alkaline conditions (**Figure 1**). To reduce the leachability of radioactive Cs, cement solidification should be prepared on a large scale with a value of 1 m^3 so that the specific surface area contacting water is limited. Another improvement is the use of blast furnace cement rather than ordinary Portland cement. Since cement includes Cr as a material component, solidified pieces can potentially leach hexavalent chromium.

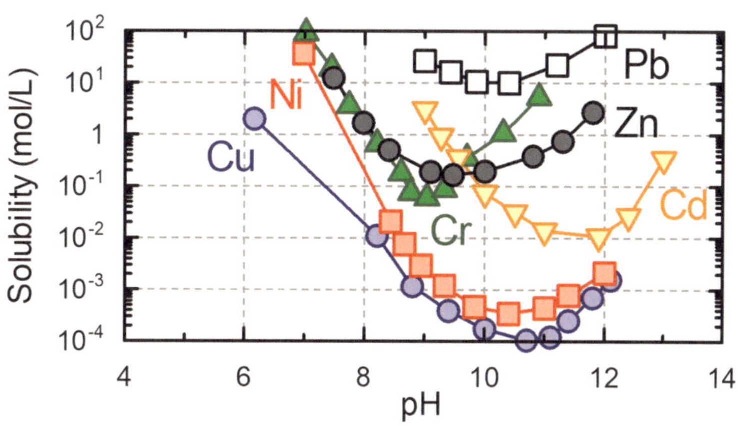

Figure 1.
Relationship between solubility and pH for heavy metal [11].

Blast furnace cement can provide a reductive condition, resulting in the transformation from hexavalent to trivalent chromium leaching. In addition, blast furnace cement is effective for reducing volume expansion and maintaining the long-term containment performance of cement solidifications. APC residue complicates the solidification mechanism of cement because of coexisting reactive chemicals, such as calcium, aluminum, and sulfur. Their chemical components contribute to form an expansive mineral known as ettringite. Volume expansion due to ettringite will generate cracks on the surfaces of solidified pieces, thus increasing the specific surface area. The reduction of both specific surface area and volume expansion is an essential design criterion for controlling the cement solidification of hazardous wastes.

3.2 Solidification by magnesium oxide

Magnesium oxide (MgO) is also an effective binder for solidifying wastes containing heavy metals. There are two methods of producing MgO. One is to bake the natural magnesium carbonate included in dolomite and then crush it. The other is to precipitate Mg ions in seawater as hydroxides and then dehydrate them at high temperatures. This means that magnesium oxide is a safe insolubilizer free from toxic chemicals, which is remarkably different from cement-containing Cr.

When MgO is dissolved in water, magnesium hydroxide is precipitated so that the pH reaches around 10.5 at an equilibrium state. At this pH level, some heavy metals can exhibit the lowest solubility, as shown in **Figure 1**. This has been considered a reason why MgO has a greater ability than cement to immobilize heavy metals. On the other hand, MgO cannot give such a large compressive strength to solidified pieces compared with cement, and MgO is about 8–10 times more expensive than cement. Therefore, it is necessary to optimize the amount and field of usage of MgO. **Figure 2** shows the results of batch leaching tests using an APC residue solidified with blast furnace cement or magnesium oxide. The ratio of ash (A) to binder (B) in weight is parametrically changed. Solidification remarkably has reduced the leachability of Cd, Zn, and F compared with raw APC residue. However, the leachability of Pb cannot be reduced by cement solidification even by increasing the amount of cement. In contrast, magnesium oxide can reduce the leaching amount under specified conditions.

3.3 Stabilization/solidification of mercury-containing waste

According to the framework of the Minamata Convention, the national scheme for the appropriate disposal of Hg-containing waste is required. Due to its environmental effects, Hg-containing waste must be stabilized prior to disposal in HgS form and/or solidified with a polymer or cement to reduce leaching and volatilization. **Figures 3** and **4** show the long-term leaching and volatilization behaviors of processed mercury. Hg-containing waste specimens are first stabilized with sulfide as metacinnabar [13, 14], which has extremely low solubility in water. The specimens are then solidified with one of four binders—sulfur polymer (SP), low-alkaline cement A (CA), low-alkaline cement B (CB), or low-alkaline cement B containing a water-reducing agent (CB+). Here, low-alkaline cement A has hauynite as the main component, and low-alkaline cement B has high-volume fly ash and silica fume. Stabilized Hg-consisting waste solidified with a sulfur polymer exhibits the lowest leaching and volatilization. Low-alkaline-cement-based binders effectively confine Hg but have lower performance than sulfur polymer. pH may significantly affect Hg leachability [15]. Leaching from a piece solidified by low-alkaline-cement binders increases under acidic

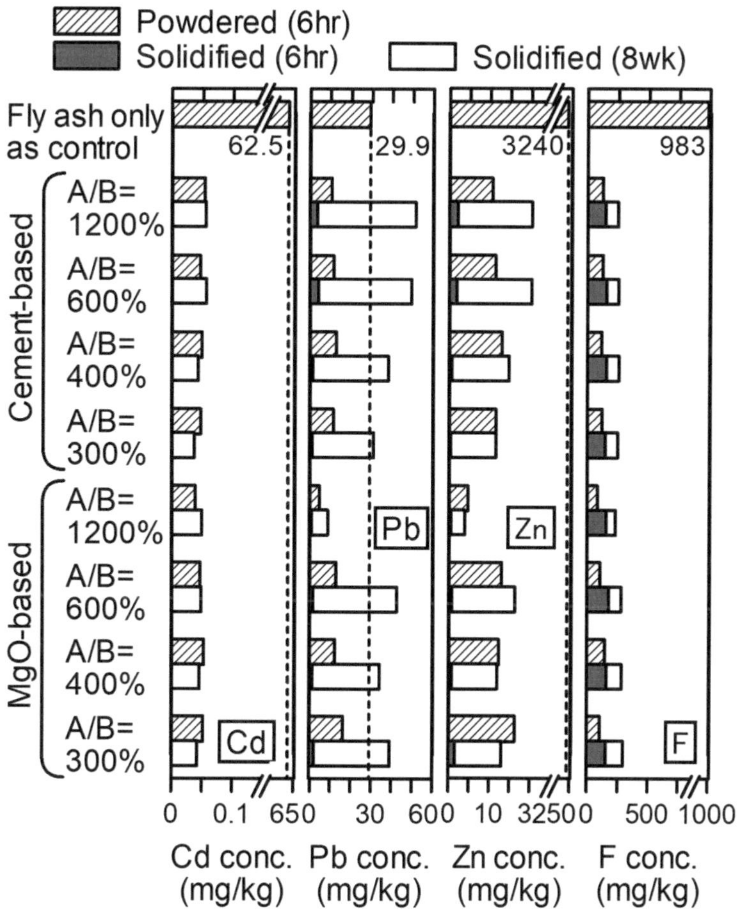

Figure 2.
Results of batch leaching tests using an APC residue solidified with blast furnace cement or magnesium oxide [12].

conditions, whereas that solidified with sulfur polymer increases under alkaline conditions. On the other hand, Hg volatilization increases with temperature except for waste solidified with sulfur polymer. Sulfur polymer is effective for decreasing the volatilization rate due to elevated temperature.

3.4 Stabilization by diatomite addition

Among available pretreatments, cement-based S/S is commonly used worldwide [16, 17]. In this process, calcium–silicate–hydrate (C–S–H) gel is formed by the reaction between amorphous silica ($SiO_2{\cdot}nH_2O$) and calcium hydroxide ($Ca(OH)_2$) (pozzolanic reaction) in the cement [18]. Toxic metals can be immobilized by the C–S–H gel via sorption, incorporation, and encapsulation owing to the high microporosity and high surface area [19, 20]. APC residues usually contain high amounts of Ca as a sorbent and reactant for the removal of acidic components in exhaust gas [21]. APC residues often show high pH due to the presence of alkaline Ca compounds [22, 23], and the solubility of amorphous silica increases at alkaline pH [24]. Owing to the high Ca content and alkaline pH provided by APC residues, the addition of amorphous silica to APC residues may induce C–S–H gel formation via pozzolanic reactions for

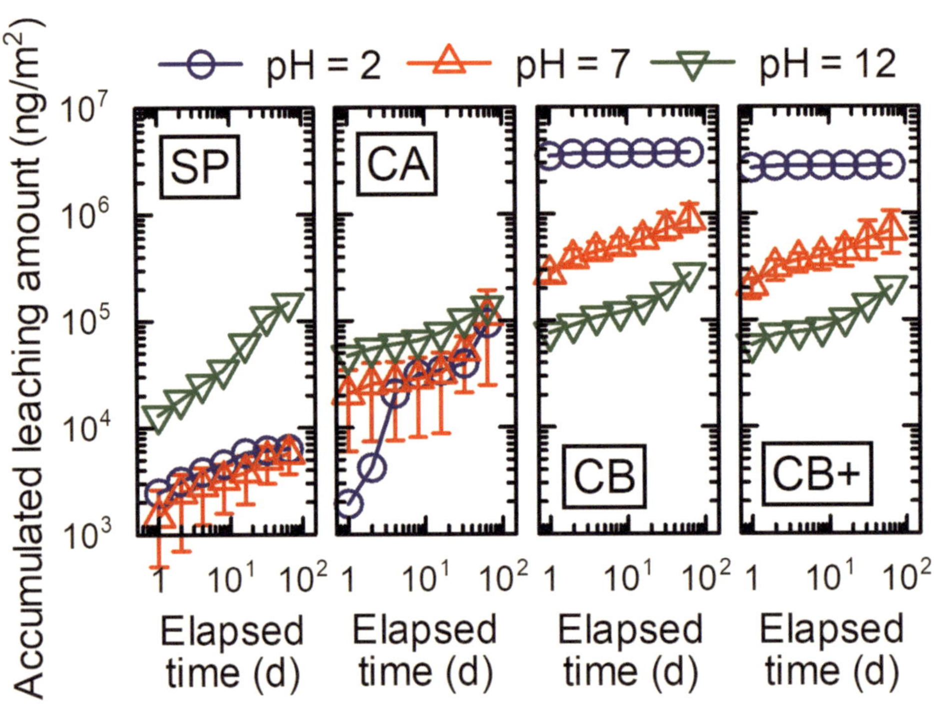

Figure 3.
Results of long-term leaching tests: Effects of pH on Hg leaching.

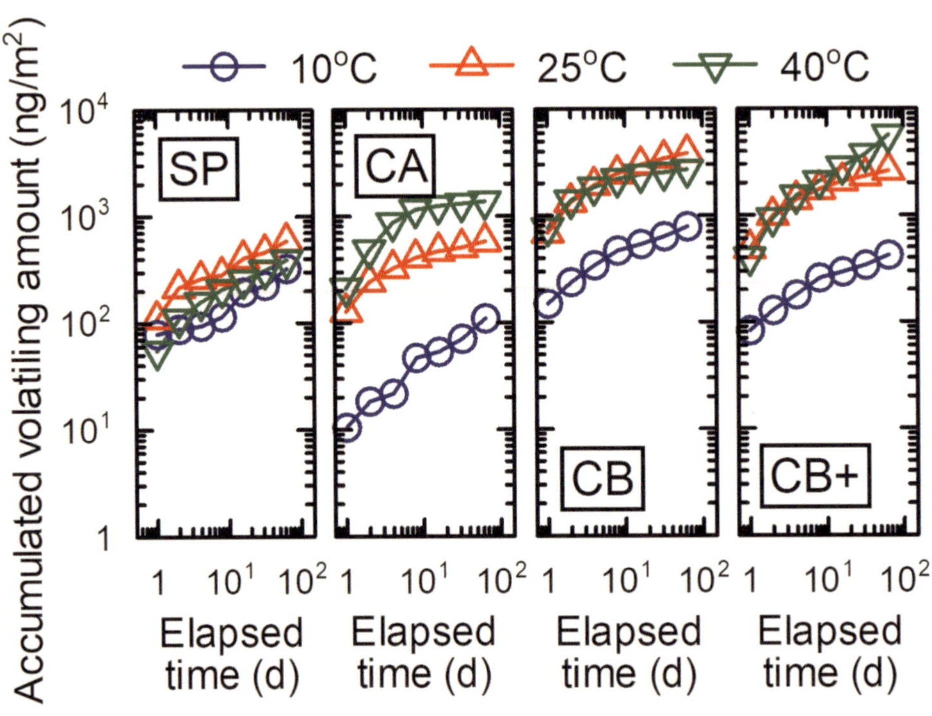

Figure 4.
Results of long-term volatilization tests: Effects of temperature on the accumulated amount of volatilized Hg.

metal immobilization. Thus, both the treatment cost and the use of chemical agents might be reduced by using inexpensive silicon materials instead of cement. We considered diatomite as a natural pozzolanic additive [25] for lead immobilization in APC residues owing to its high amorphous silica content [26, 27], relative abundance [28], and low cost compared to Portland cement [29].

Assessment of the impact of diatomite addition on Pb immobilization in APC residues (**Figure 5**) indicates that Pb leaching from weathered APC residues decreased as time and temperature increased. This is attributed to the increase in the hydration reaction of cementitious materials as the temperature increases [30, 31]. At each weathering temperature, Pb leaching from stabilized APC residues decreased as diatomite doses increased. The leaching amount of Pb from 14-day stabilized APC residues with 0%, 5%, or 10% diatomite addition was reduced by 18–67%, 67–90%, or 80–99%, respectively. Consequently, the leaching amount of Pb dropped below 0.3 mg/L (Japanese criterion for landfill disposal) after 14 days of curing with the addition of 10% diatomite at 70°C.

Figure 6 shows the X-ray diffraction (XRD) patterns of raw APC residues and 14-day cured APC residues following the addition of 10% diatomite at 70°C. The peak

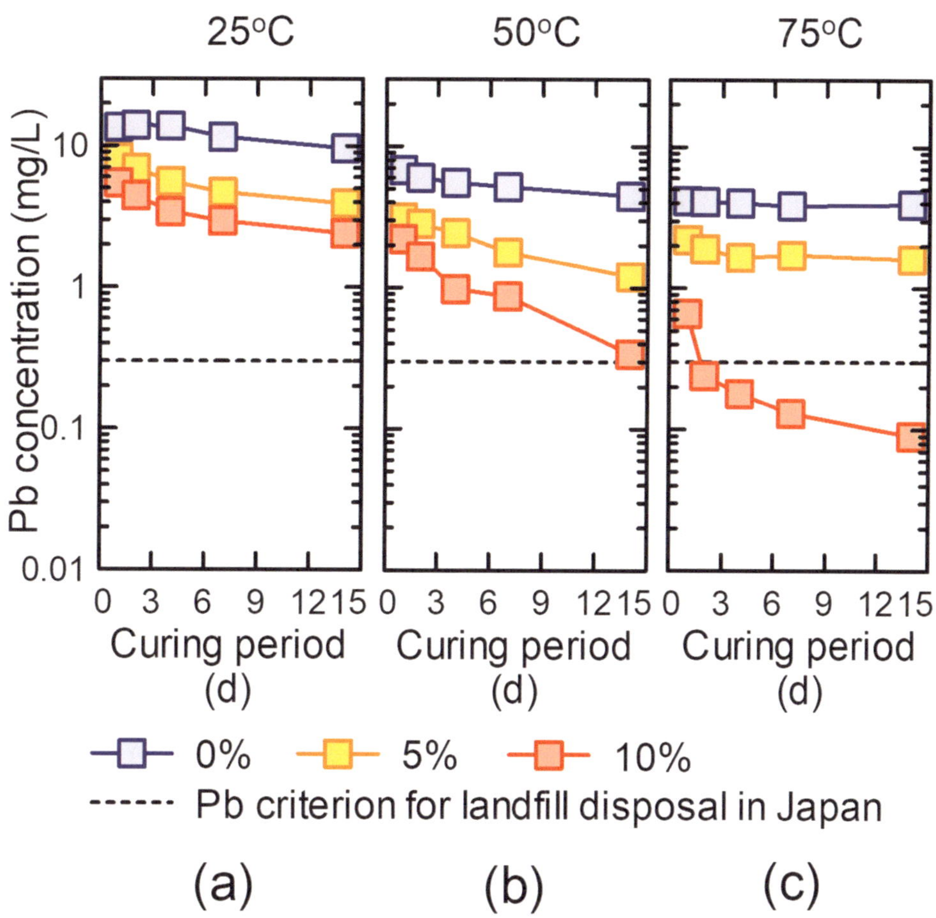

Figure 5.
Leaching concentrations of Pb from cured APC residues. APC residue under the temperature of (a) 25°C, (b) 50°C, (c) 75°C.

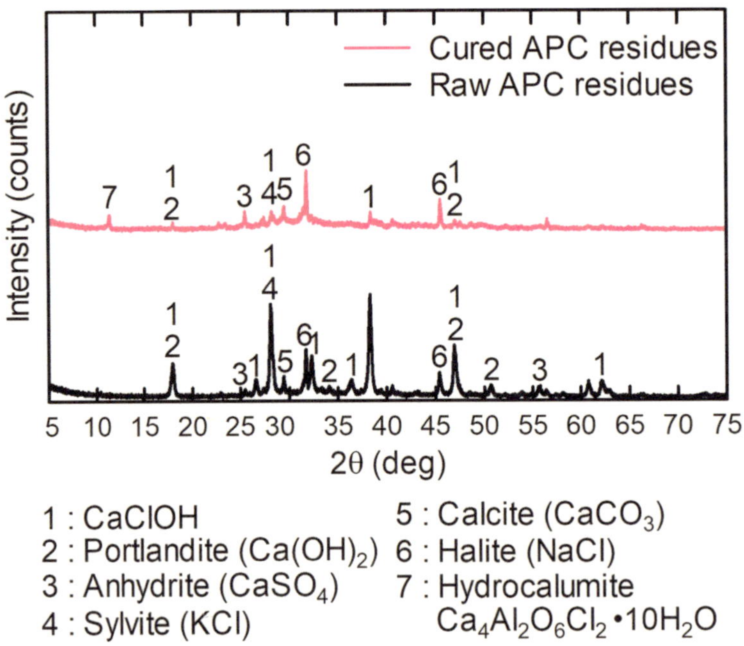

Figure 6.
XRD patterns of raw and cured APC residues by 10% of diatomite at 70°C for 14 days.

intensities of CaClOH and $Ca(OH)_2$ significantly decreased by weathering with diatomite, indicating that they reacted to diatomite and were consumed by the **Figure 6** XRD patterns of raw and cured APC residues by 10% of diatomite at 70°C for 14 days pozzolanic reaction. New peaks of C–S–H gel in the stabilized APC residues were not confirmed, as the residues were below the detectable level of XRD analysis or crystallization was incomplete [32]. Even though C–S–H gel formation was not detected in the XRD analysis, the amount was sufficient to immobilize 99% Pb in the APC residues.

Diatomite, consisting mainly of amorphous silica, was used as a pozzolanic additive for Pb immobilization in APC residues instead of cement. The results showed that the leaching amount of Pb from the stabilized APC residues was reduced by C–S–H gel formation via the pozzolanic reaction among $Ca(OH)_2$, CaClOH, and diatomite. Consequently, the leaching amount of Pb dropped below 0.3 mg/L. This study showed the feasibility of using diatomite to immobilize Pb in APC residues. Although 10% diatomite was added to the APC residues, the volume increase is supposed to be lower than that in cement-based S/S [33]. From the viewpoint of landfill management, this treatment would reduce the use of chelating agents while suppressing the increase in volume. If wastes containing amorphous silica can be used to immobilize metals in APC residues, this method has the potential to be a low cost and environmentally friendly solution.

4. Weathering and attenuation of waste in landfills: impact of mineral compositions on metal leachability

Waste containing toxic metals, which are commonly immobilized before disposal, will be stabilized by a chemical or microbial weathering process in landfills. For example, precipitation of insoluble metal sulfide in the presence of sulfides produced by sulfate-reducing bacteria is one of the major mechanisms by which toxic metals are immobilized in landfills [34]. In the initial stage of the weathering process, calcite ($CaCO_3$) generated by carbonation via atmospheric CO_2 immobilizes toxic metals by absorption on the surface and/or incorporation into the crystal structure [35, 36]. Thus, these processes have non-negligible impacts on metal mobilities at waste landfills. In this context, the impacts of mineral compositions of waste in a landfill on the leaching behaviors of heavy metals were analyzed. The obtained waste was composed of the following—cover soils (0–0.15 m), waste layers (0.15–47.6 m), embankment (47.6–50.9 m), and gravel layers (50.9 m–56 m). Mineral compositions identified by XRD analysis (**Table 1**) showed that the waste in this landfill mainly consisted of calcite ($CaCO_3$), gypsum ($CaSO_4{\cdot}2H_2O$), and quartz (SiO_2). The contents of heavy metals (Cu, Mn, Ni, and Zn) and their leachability (**Figure 7**) revealed that the waste at 31.3 m depth had high contents and high leaching concentrations of Cu, Mn, Ni, and Zn. The core sample contained weddellite ($CaC_2O_4{\cdot}2H_2O$) and whewellite ($CaC_2O_4{\cdot}H_2O$). Waste incineration residues are known to contain small amounts of organic acids, such as oxalic acid ($C_2H_2O_4$) [37]. Calcium oxalate (weddellite and/or whewellite) can be formed in waste incineration residues by the coprecipitation of Ca^{2+} ion and oxalic acid [36]. It is also considered that calcium oxalate in sludge produced by the neutralization of waste acid using calcium hydroxide has been landfilled. Another possible source of calcium oxalate is waste from the ceramics industry. The waste also contains iron oxide (see **Table 1**) that, like calcium oxalate, is used for ceramic glazes. Thus, oxalic acid and calcium oxalate could be in the waste. If oxalic acid has a high capability of extracting heavy metals, it might enhance the leachability of heavy metals in waste landfills. Compared to mineral acids, the contents of oxalic acid (organic acid) do not seem high, owing to its precipitation as a metal–oxalate complex [22]. On the other hand, the mobility of heavy metals seems to be significant in this landfill according to the solubility of the metal–oxalate complex (**Table 2**) [38]. Moreover, the stability constant of this complex is higher than that of calcium oxalate (**Table 2**), suggesting that the mobility of heavy metals might be enhanced by the

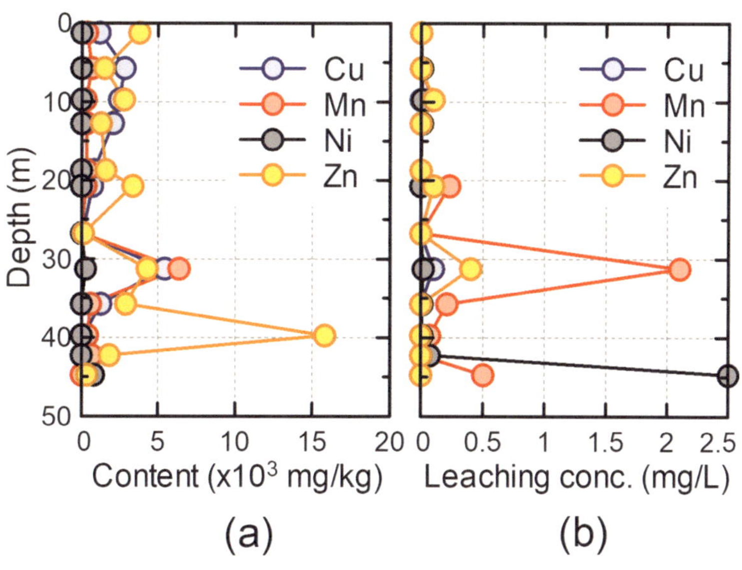

Figure 7.
Contents and leaching concentrations of heavy metals in waste. (a) contents, (b) leaching concentrations.

Depth (m)	Albite	Calcite	Chlorite	Gibbsite	Gypsum	Iron oxide	Muscovite	Quartz	Weddellite	Whewellite
1.25		+			+			+		
5.75		+			+			+		
9.75		+			+			+		
12.75		+			+			+		
18.75		+						+		
20.75		+			+			+		
26.75		+						+		
31.25		+			+	+		+	+	+
35.75		+						+	+	+
39.75	+	+	+				+	+		
42.25		+			+			+		
44.75				+	+			+		

+: Presence

Table 1.
Mineral compositions of a waste obtained from the landfill.

DOI: http://dx.doi.org/10.5772/intechopen.102872

Metal	Ca^{2+}	Cu^{2+}	Zn^{2+}	Pb^{2+}
Solubility in water (mg/100 mL)	0.67	2.53	0.79	0.16
Stability constant	3.00	6.25	4.87	4.91

Table 2.
Solubility and stability constant of metal oxalate in water.

formation of a metal–oxalate complex via a substitution reaction between Ca^{2+} ion in calcium oxalate and divalent metals.

Our assessment revealed the high leachability of Cu, Mn, Ni, and Zn from waste containing calcium oxalate. Calcium oxalate could be in waste from possible sources, such as incineration residues, sludge produced by waste acid treatment, and ceramics. This suggests that the existence of specific minerals, such as calcium oxalate (weddellite and/or whewellite), might enhance the leachability or mobility of heavy metals in landfills to some extent.

5. Soil adsorption barrier for retarding transport of heavy metals

5.1 Distribution coefficients

The soil adsorption performance of heavy metals, which directly affect migration, has been of great interest among engineers. Many reports about adsorption parameters for various soil types have been published. In particular, the Japan Atomic Energy Agency (JAEA) has established a database summarizing the results of soil adsorption tests using radioactive isotopes [39]. This section overviews soil adsorption parameters (distribution coefficients) of radioactive isotopes, such as Hg, Cd, Pb, Se, and Cs, as references. Cr and As do not have radioactive isotopes and JAEA does not support their distribution coefficients. The distribution coefficients are collected with previous references targeted to soils in major countries (**Figures 8–14**) [40–55].

The distribution coefficients of any heavy metals take a wide range of values. Here, the geometric mean value in each soil type against heavy metals is presented as a representative value in **Figures 8–14**. The distribution coefficient indicates a soil

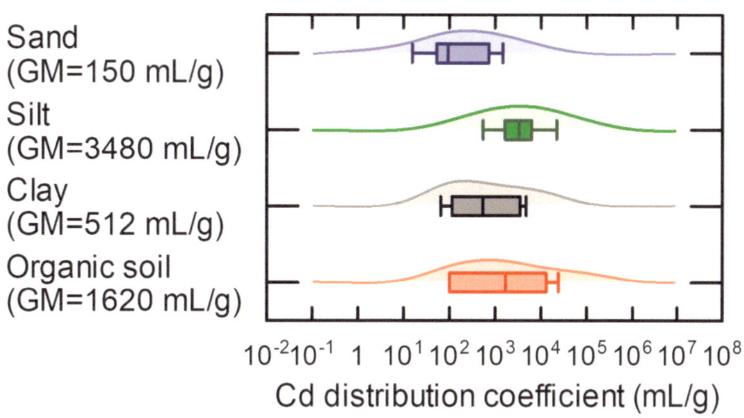

Figure 8.
Distribution coefficients of cadmium arranged from JAEA database.

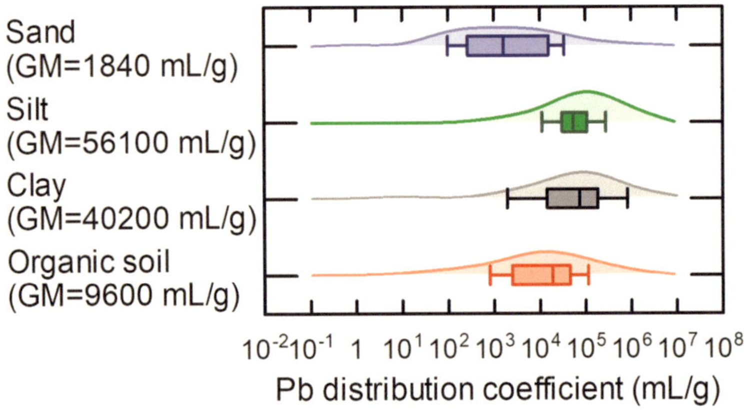

Figure 9.
Distribution coefficients of lead arranged from JAEA database.

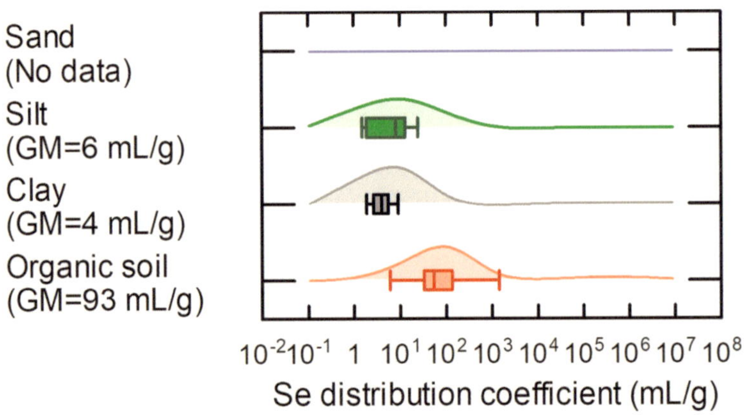

Figure 10.
Distribution coefficients of selenium arranged from JAEA database.

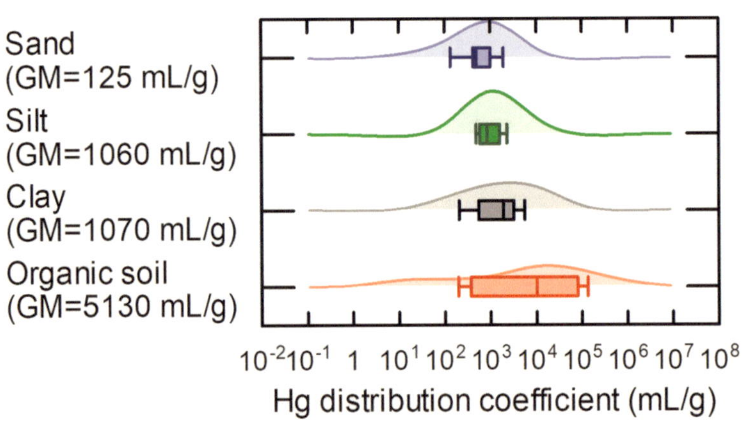

Figure 11.
Distribution coefficients of mercury arranged from JAEA database.

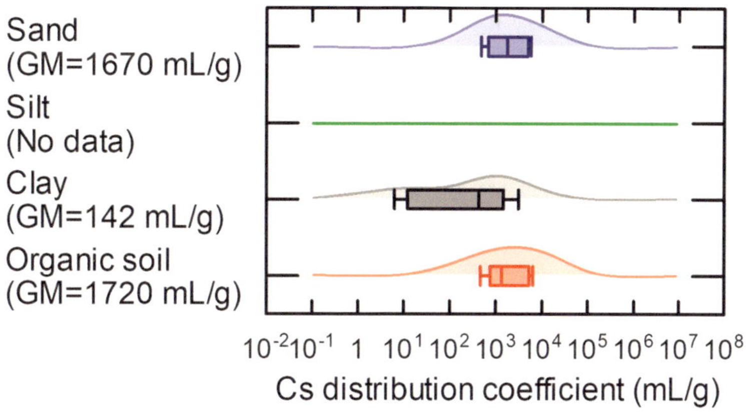

Figure 12.
Distribution coefficients of cesium arranged from JAEA database.

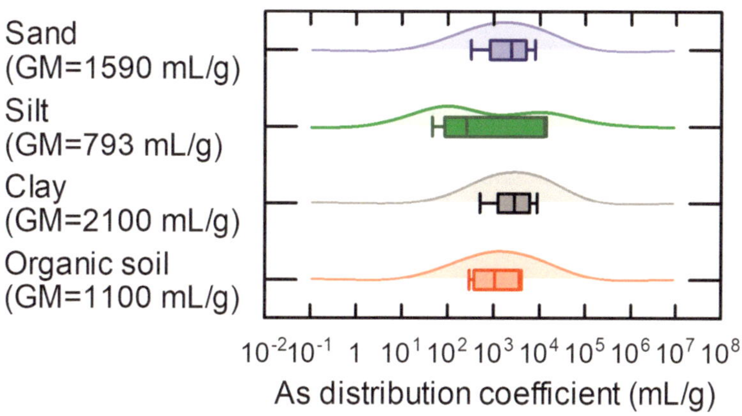

Figure 13.
Distribution coefficients of trivalent or quinquevalent arsenic.

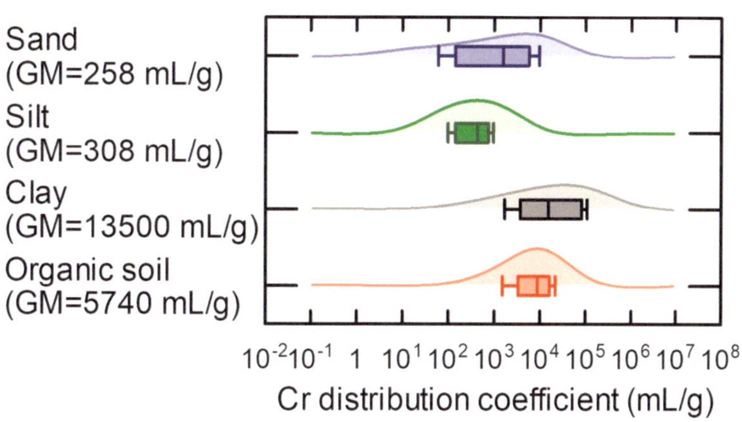

Figure 14.
Distribution coefficients of trivalent or hexavalent chromium.

type's ability to adsorb a heavy metal. The larger the distribution coefficient, the greater the retardation of the chemical substance transport, resulting in superior barrier performance against the transport. The distribution coefficients against Cd, Pb Hg, and Cs are relatively high, whereas those against Se, As, and Cr are lower. This is because the chemical forms of Se, As, and Cr in water are anions that are hard to adsorb to the soil surface with a negative charge and whose transport is hard to retard. Therefore, the environmental impacts of the transport of Se, As, and Cr with such small distribution coefficients should be carefully evaluated.

Se, As, and Cr in water can have anionic forms with different ionic valences, depending on environmental conditions, such as pH and oxidation–reduction potential. Numerous review studies, including this chapter, describe in broad strokes the distribution coefficients of heavy metals, but they hardly investigate the differences in distribution coefficients of heavy metals with different ionic valences. However, some scientific papers investigate the effects of ionic valence on distribution coefficients. For example, hexavalent chromium has smaller distribution coefficients than trivalent chromium [54], but both have values in the range of 100 mL/g or less at maximum [55]. Kumpiene et al. [56] review the stabilization mechanisms of As, Cr, Cu, Pb, and Zn. Especially, they discuss the stabilization of As, which is dependent on seven factors—iron compounds, aluminum oxides, manganese oxides, organic matter, alkaline materials, clay minerals, and sulfides. The fact that As can be adsorbed on Fe has been considered a reason why As and Pb included in slags do not leach into water [57–59].

Soil pollution is a global environmental problem. As and Cr are relatively common as causative pollutants, so, their findings are collected and shared among not only researchers but also practitioners. In contrast, Se is a relatively minor substance in soil pollution and waste management, and thus far there are few studies on Se. Further studies on Se, As, and Cr are needed to accurately manage human risk because their distribution coefficients are small and environmental pollution by them is easily spread.

5.2 Soil adsorption barrier for gaseous Hg

Soil adsorption is also effective for retarding gaseous substances and preventing their diffusion. In general, there are three methods for evaluating the distribution coefficients of gaseous substances—(i) dynamic adsorption column, (ii) gravimetry, and (iii) constant volume [60]. The constant volume method has commonly been used to obtain adsorption isotherms. However, gases often adsorb on the container surface or leak from the plug, so obtained isotherm data need to be compensated using the losses of the gases in a blank test. In the previous studies that evaluate the adsorption abilities of adsorbents against some volatile organic compounds (VOCs) using Tedlar bags [61], the compensated isotherms at the equilibrium state can be exactly calculated because the losses of the VOCs in the Tedlar bags are mostly due to adsorption on the surface.

Gaseous Hg, however, would not only adsorb on the container surface but also leak from the plug, so, the equilibrium state cannot be reached. This characteristic of gaseous Hg makes the evaluation of its distribution coefficients difficult. Therefore, a testing method to evaluate adsorption abilities under a nonequilibrium state caused by the leakage should be established.

Ishimori et al. [62] suggested the constant volume method for evaluating the adsorption characteristics of soils and adsorbents against gaseous Hg under a nonequilibrium state. They formulated the phenomenon of nonequilibrium soil adsorption with leakage using the Langmuir sorption model and the diffusive leakage

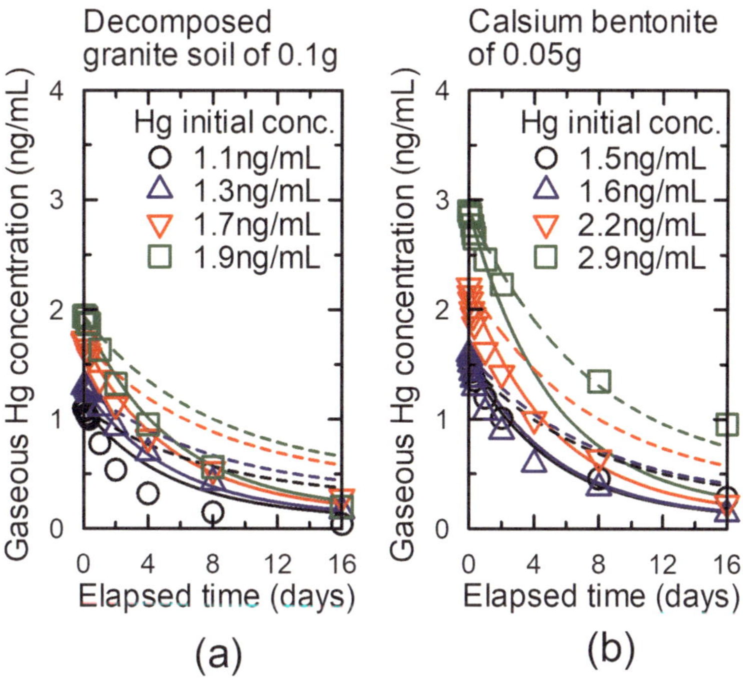

Figure 15.
Kinetics of gaseous Hg during adsorption tests for decomposed granite soil (a) and calcium bentonite (b). Plots: Experimental results, solid lines: Fitting results to a governing equation, dashed lines: Estimation results with neither Hg adsorption on container surface nor leakage from sealing plug.

model, resulting in an estimation of distribution coefficients of gaseous mercury by fitting experimental data to the governing equation (see **Figure 15**). Finally, the adsorption isotherms for sand, granite soil, calcium bentonite, and mordenite are estimated as shown in **Figure 16**. Then their distribution coefficients values for gaseous mercury are obtained as 56.3, 2070, 7140, and 3490 mL/g, respectively. It is noted that their values are obtained from the initial slopes of their adsorption isotherms when the equilibrium concentrations are zero.

Hg-coning wastes will be disposed of in landfill sites in the near future due to the signing of the Minamata Convention on Mercury. Soil adsorption barriers are an effective containment method to retard the transport and minimize the emission of mercury. The distribution coefficient is the most important parameter for providing the required containment barrier performance in landfills. The soil adsorption characteristics against both the aqueous and gaseous forms of mercury are insufficiently investigated thus far. It is well known that distribution coefficients depend not only on the type of soil and the adsorbents but also on environmental conditions, such as pH, oxidation–reduction potential, temperature, and coexisting aqueous or gaseous substances.

6. Long-term environmental safety evaluations using multiphysics numerical simulations

Transport of aqueous and gaseous Hg in controlled landfills.

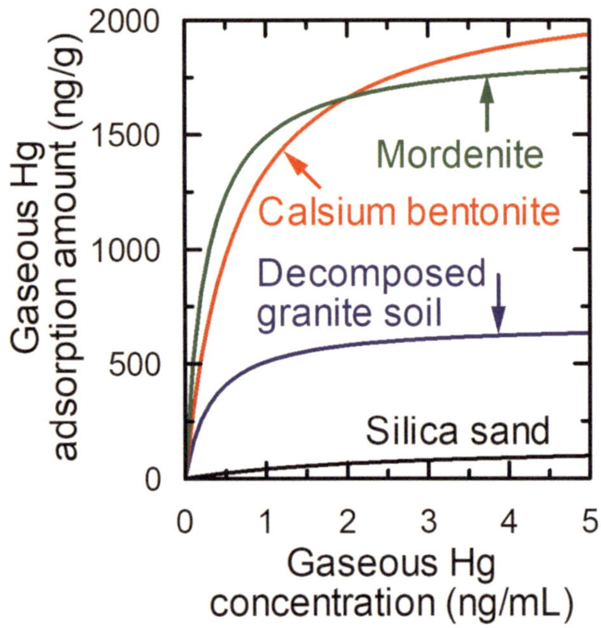

Figure 16.
Adsorption isotherms of four samples against gaseous Hg [59].

In environmental engineering, numerical simulation is a method of predicting the fate and transport of chemicals. It is especially effective for evaluating long-term environmental safety, which cannot be predicted in experiments under limited conditions, such as testing duration, the scale of the domain targeted, and its heterogeneity. Numerical simulation can be conducted by solving governing equations that mathematically express the phenomenon targeted by prediction. Governing equations are always formulated to satisfy the law of mass conservation for targeted substances. Optionally, governing equations are also formulated to satisfy the law of momentum or energy conservation. Here, examples of numerical simulations for the transport of aqueous and gaseous mercury in landfills after disposal of mercury-containing waste are introduced, and the effectiveness of those simulations for long-term safety evaluation are discussed.

Ishimori et al. [63] investigated the environmental safety of landfill sites in which Hg-consisting waste is disposed of. Serial batch tests are conducted to evaluate the long-term leaching and volatilization of Hg stabilized in its sulfide form and solidified using either a sulfur polymer or low-alkaline cement. Using measured Hg leaching and volatilization rates, numerical simulations are conducted to investigate the long-term behavior of Hg after its disposal in landfill sites.

Figure 17(a) shows the analysis domain and conditions. A landfill site disposing of Hg-containing waste is modeled in the cross-sectional domain, and the waste is treated through stabilization and solidification techniques as shown in **Figures 3** and **4**. After treatment, the waste takes the form of a 1 m^3 cube. The entire array of solidified cubes is covered with a soil sorption layer, designed to retard the transport of emitted Hg. A drainage pipe is placed at the bottom of the analysis domain to accumulate the leachate, and a final cover or cutoff layer to reduce rainfall permeation is placed on top. **Figure 17(b)** and **(c)** show the initial and boundary conditions of the analysis—the boundary condition at the top of the domain indicates the rainfall

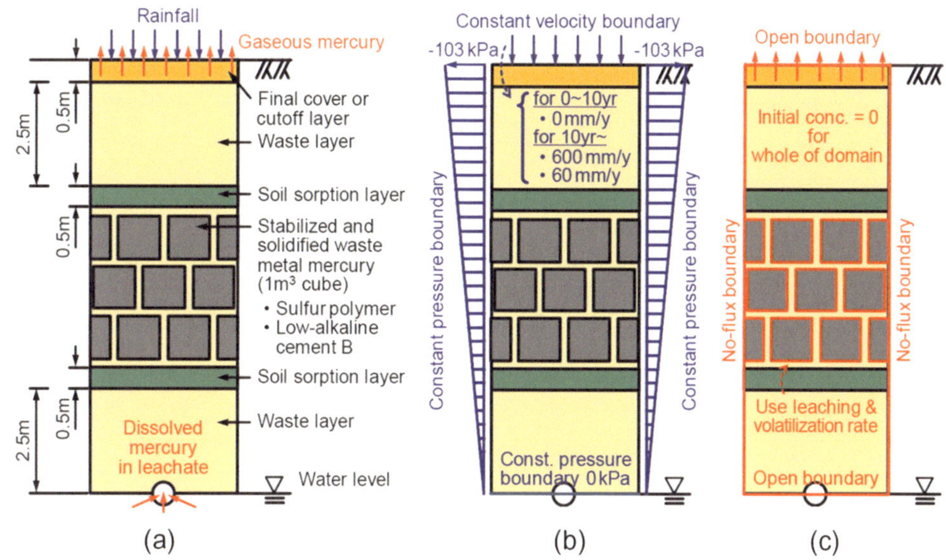

Figure 17.
Analysis domain and conditions: (a) two-dimensional cross section of the landfill, (b) initial and boundary conditions for seepage analysis, (c) initial and boundary conditions for advection–dispersion analysis [60].

intensity. For the first 10 years of the analysis, the rainfall intensity for the top boundary condition is considered to be 0 mm/y, as landfilling of the solidified piece would be carried out under a roof; at the end of the landfilling process, the roof would be removed, so in subsequent years, rainfall at the top boundary condition is considered to permeate the final cover with an intensity of either 600 mm/y or 60 mm/y. These different rainfall intensity values are used to evaluate the effects of using a cutoff layer covering the waste site, which would decrease the overall ingress of water to the landfill site. The measured leaching and volatilization rates of mercury are applied to the surface boundaries of the stabilized, solidified Hg-consisting waste.

This numerical simulation investigates the effects of the soil sorption and cutoff layers on the concentrations of dissolved Hg in the leachate at the bottom of the drainage pile and of gaseous Hg emitted from the final cover—the analytical conditions for these models are listed in **Table 3**. These parameters are given to the governing equations. In this study, they are formulated based on the law of mass conservation regarding water, air, dissolved Hg, and gaseous Hg, where the governing equations for water and air are called two-phase flow models, and the governing equations for dissolved and gaseous Hg are called advection–dispersion equations in general. Their equations have been widely used in numerical simulations for fluid flow and chemical substance transport in porous media. A notable point is that the phase transfer rate between dissolved and gaseous Hg is modeled using the Henry constant of Hg. The equation systems mentioned above can be numerically solved using many commercial simulation software programs or open-source codes. The following results are obtained from numerical solutions by COMSOL Multiphysics ver 5.0 (COMSOL, Inc).

A drastic difference between solidification by sulfur polymer and that by low-alkaline cement appears in the total Hg emissions from the landfill (**Figure 18**). The total Hg emission depends significantly on the presence of a soil sorption layer and

Parameters	Unit	Waste layer	Soil sorption layer
Porosity	1	0.3	0.3
Intrinsic permeability	m^2	1×10^{-12}	1×10^{-12}
Dry bulk density	kg/m^3	1400	1400
VG parameter, α	1/m	2	2
VG parameter, n	1	1	1
Longitudinal dispersivity	m	3	3
Transversal dispersivity	m	1	1
Distribution coefficient	mL/g	0	100
Henry constant	1	0.43	0.43

Table 3.
Analytical conditions for predicting mercury behavior in landfill.

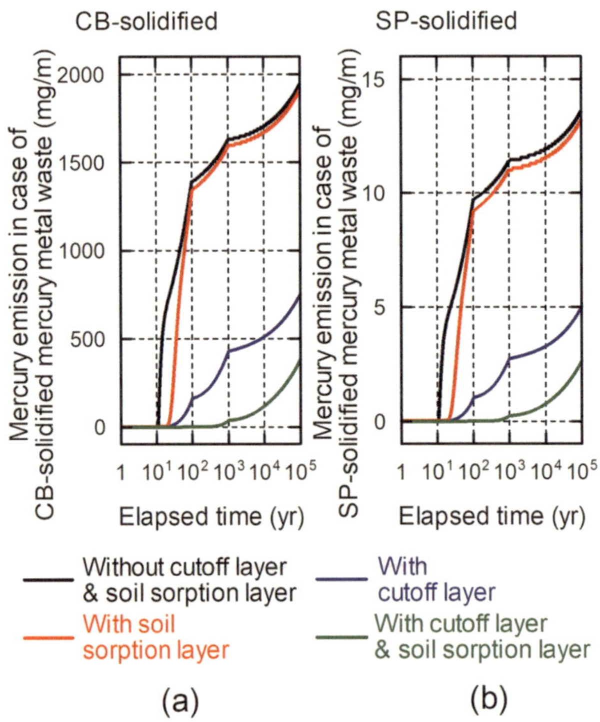

Figure 18.
Total amount of mercury emitted from a landfill with Hg-consisting waste solidified by (a) sulfur polymer or (b) low-alkaline cement.

cutoff layer as well as on types of binders to solidify the Hg-consisting waste. The most effective measure to reduce Hg emission is considered to be sandwiching sulfur polymer-solidified pieces between sorption layers and covering the landfill surface with a cutoff wall. Numerical simulations will help us design the required geometry and material quality of the soil sorption layer, the cutoff wall, and the stabilized solidified Hg-consisting waste.

Hazardous waste containment performance depends on the aging of RC materials or their failure due to chemical attack.

Another practice of numerical simulation is the multiphysics of seismic analysis and reactive chemical transport analysis to evaluate the long-term environmental safety of isolation-type landfills that are designed for hazardous waste and that have a reinforced concrete structure [64]. Hazardous wastes possess a hazard to human health and the environment when improperly managed. They have extremely high leaching concentrations, so they cannot be disposed of directly into regular landfill sites. In general, such hazardous wastes are strictly controlled in waste containment facilities whose function is to prevent penetration and thus to avoid the wastes from leaching due to rainfall [2, 3]. As case studies, this containment facility so-called isolation-type landfills have been built using waterproof reinforced concrete with a thickness > 350 mm and compressive strength >25 MPa, based on the regulations in Japan. Ishimori et al. [65] used numerical simulations to show the importance of a multi-barrier system consisting of stabilization/solidification techniques and artificial/natural soil sorption layers to minimize the negative impacts of a hazardous waste landfill. These numerical simulations consist of seismic analysis to evaluate the stability of the structure in the event of huge earthquakes and reactive chemical transport analysis to predict the long-term leaching concentration profiles from landfills damaged due to deterioration over time or sudden huge earthquakes.

Figure 19 overviews those numerical studies. The analysis domain consists of a waste containment facility and its surrounding grounds. The environmental safety of hazardous waste in the facility, which is built with reinforced concrete, is evaluated by predicting the concentrations of heavy metals at monitoring well located in the lower reaches of the groundwater. Four numerical studies are performed. In Case 1, no earthquake occurs for 100 years. In Cases 2 and 3, a small and a large earthquake, respectively, occur after 5 years. Each earthquake is equivalent to a magnitude of a Level 1 (L1)

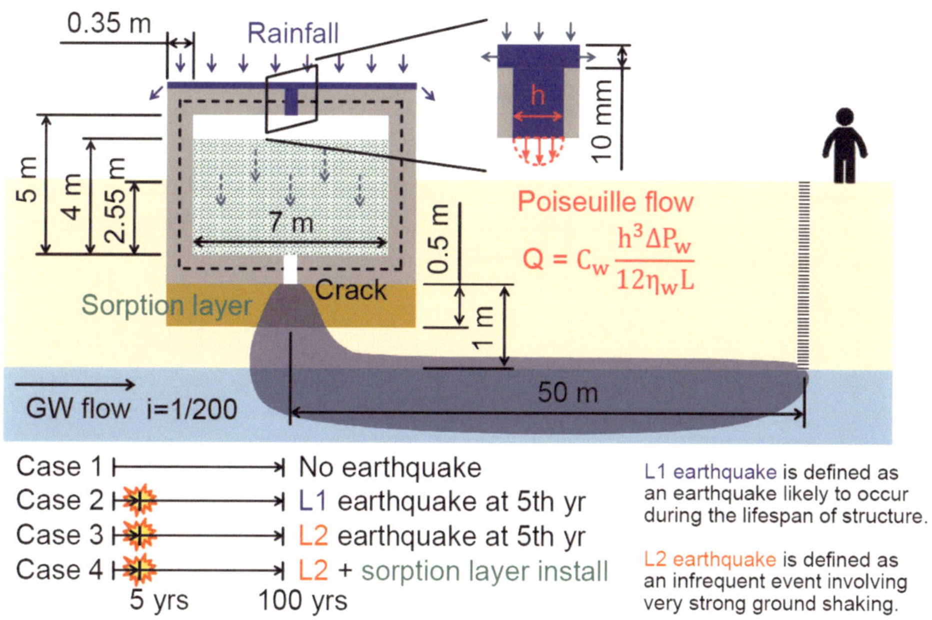

Figure 19.
Analysis domain and conditions for case studies.

or Level 2 (L2) earthquake as defined in Japan, respectively. Case 4 additionally has a 50-cm sorption layer underlying the waste containment facility. Hazardous waste is assumed to be APC residue, and cadmium is targeted as a contaminant.

Figures 20–22 overview the whole analytical procedure. **Table 4** shows the main analytical conditions used. The governing equations are formulated from the law of

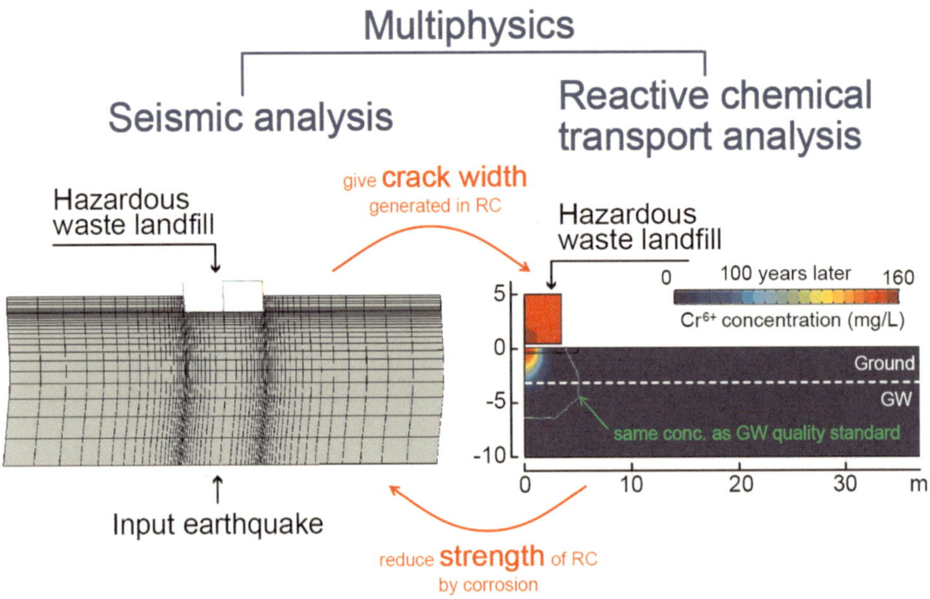

Figure 20.
Multiphysics.

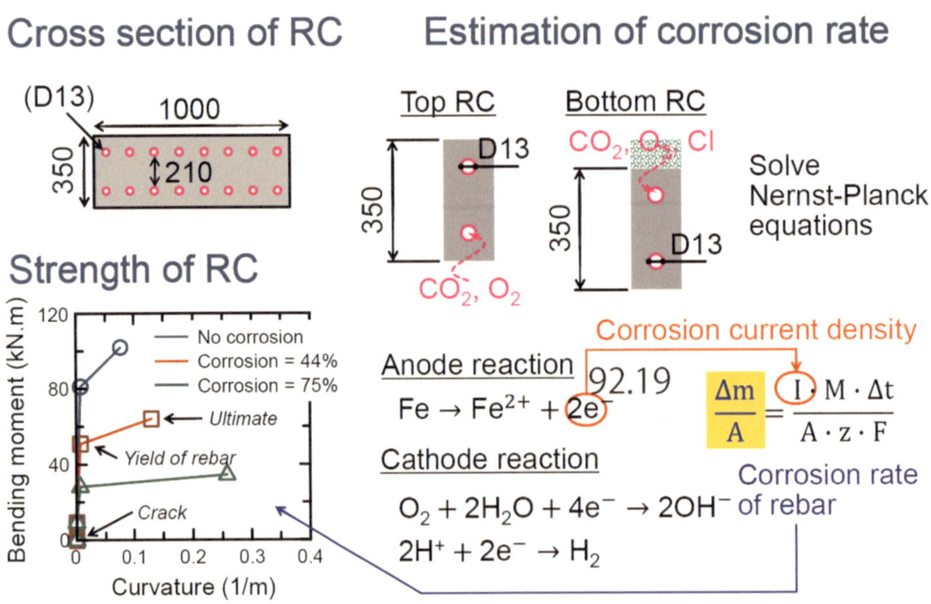

Figure 21.
Concept of how to reduce RC strength; from structure analysis given reduced strength, axial force, 2 and bending moment of RC beams are calculated.

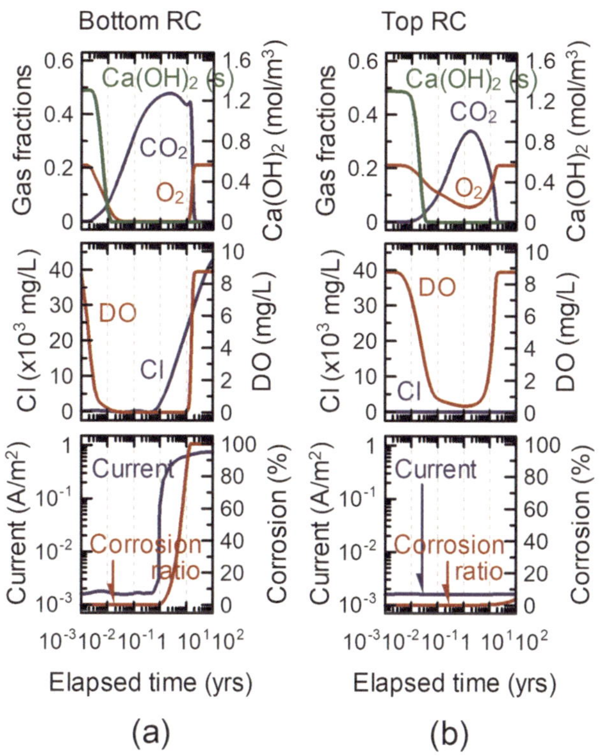

Figure 22.
Concentration profiles in the containment facility; (a) bottom RC, (b) top RC.

Parameter	Value
N-values for ground assuming sandy loam, N	30
Shear wave velocity, V_s	249 m/s
Maximum oxygen consumption rate, $\lambda_{O2\,max}$	4.9×10 d^{-2} d^{-1}
First-order decay rate in aerobic condition, k^*	2.0 $yr.^{-1}$
First-order decay rate in anaerobic condition, K	0.2 $yr.^{-1}$
Second-order reaction rate for carbonation, λ	0.6 m^3/mol/d
Intrinsic permeability for ground, K	1.0×10^{-12} m^2
Distribution coefficient for sorption layer, K_d	300 mL/g

Table 4.
Main analytical conditions.

momentum balance to perform ground motion analysis and structure analysis. In the ground motion analysis, a governing equation having an unknown variable of horizontal displacement is solved using a given seismic wave as a boundary condition on the base layer. Then the calculated horizontal displacement and shear stress are applied as external forces acting on the landfill. Finally, a structural analysis for RC-based beams simulating the landfill geometry is conducted. From the calculated axial forces and bending moments in the beams, the crack width is estimated according to previous experimental studies [66–69]. The estimated crack width is used to

		Bending moment (kN.m)	Axial force (kN)	Failure mode	Crack width (mm)	Flow rate (mL/d/m)
Case 1	Bottom RC	6.6	39.9	No crack	0	Close
	Top RC	23.2	11.7	Crack	0.330	2900
Case 2	Bottom RC	6.2	37.9	No crack	0	Close
	Top RC	20.9	17.1	Crack	0.291	1990
Cases 3 and 4	Bottom RC	7.3	25.9	Crack	0.110	Close
	Top RC	22.2	17.9	Crack	0.306	2310

Table 5.
Results of seismic analysis.

determine the leakage rate from inside to outside the landfill according to the theory of Poiseuille flow. The governing equations mentioned above are solved using Moleman-i plus (Mizuho Information & Research Institute) for seismic analysis and COMSOL Multiphysics ver 5.0 for reactive chemical transport analysis.

Table 5 shows the results of the seismic analysis. In any top cover with reinforced concretes, a crack is generated, resulting in the infiltration of rainfall into the landfill compartment. Whereas a crack in the bottom-reinforced concrete is not generated in Cases 1 and 2, deterioration due to carbonation and salt damage is considered. The crack width in Cases 3 and 4 is estimated to be 0.11 mm, which is smaller than that in the concrete reinforced with a top cover consisting of reinforced concrete. **Figure 23** shows the results of groundwater flow and transport analysis for Cd leached from the waste containment facility. No Cd appears in Cases 1 and 2 because the bottom-reinforced concrete does not have a crack. However, Cd does leak from the crack in Cases 3 and 4; it is transported by groundwater flow and observed in the monitoring well. The soil sorption layer is effective for preventing groundwater contamination, and it is an important factor in controlling the long-term environmental safety of isolation-type landfills.

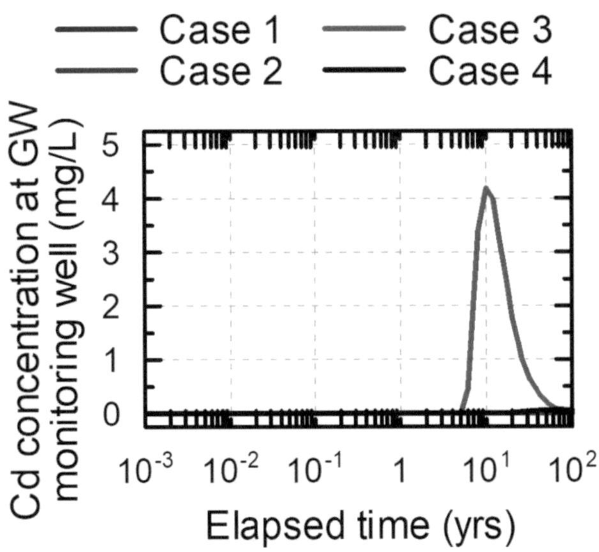

Figure 23.
Cd concentration profiles in groundwater.

This section presents a numerical simulation model to evaluate the environmental safety of hazardous waste landfills having a reinforced concrete structure. A multi-barrier system, in which a sorption layer is additionally installed under a reinforced-concrete hazardous waste landfill, can retard the transport of contaminants and will be effective for improving their environmental safety.

7. Concluding remarks

To develop an environmentally sound resource circulation society, it will be essential to implement long-term safety management practices for hazardous substances. The necessity of isolation/sequestration schemes for hazardous heavy metals is described in this chapter in terms of the multi-barrier concept, in which engineering measures, such as pretreatment and solidification technologies, control of landfill conditions, isolation barriers, and geological and artificial barriers, are implemented to attenuate environmental risks. Single artificial engineering measures should always be viewed with skepticism, and multiple barriers are necessary to retard and attenuate the migration of hazardous heavy metals. The multi-barrier concept also considers the (bio) chemical transformation behavior of heavy metals and their effect on migration. A further innovation of measures to isolate/sequester heavy metals may lead to higher levels of safety and more environmentally sound resource circulation.

Acknowledgements

This research was supported in part by the Environment Research and Technology Development Fund (3 K153004, JPMEERF18949882, and JPMEERF21357019). This research was also conducted in part under the Material Flow Innovation Research Program of the National Institute for Environmental Studies, Japan.

Conflict of interest

The authors declare no conflict of interest.

Author details

Tomonori Ishigaki*, Hiroyuki Ishimori, Hiroki Kitamura and Masato Yamada
National Institute for Environmental Studies, Tsukuba, Japan

*Address all correspondence to: ishigaki@nies.go.jp

References

[1] Brunner PH. Clean cycles and safe final sinks. Waste Management & Research. 2010;**28**:575-576. DOI: 10.1177/0734242X10370987

[2] A Working Group on Hazardous Wastes: Safe hazardous waste management systems. A state-of-the art guide for decision makers, Update 2002. Silver Springs, USA: USA/Canadian International Solid Waste Management Federation; 2002

[3] Technical Working Group of the Basel Convention. Basel Convention Technical Guidelines on Specially Engineered Landfill. Geneva, Switzerland: United Nations Environmental Programme; 2002

[4] Quina MJ, Bordado JC, Quinta-Ferreira RM. Treatment and use of air pollution control residues from MSW incineration: An overview. Waste Management. 2008;**28**:2097-2121. DOI: 10.1016/j.wasman.2007.08.030

[5] Van Der Sloot HA, Kosson DS, Hjelmar O. Characteristics, treatment and utilization of residues from municipal waste incineration. Waste Management. 2001;**21**:753-765. DOI: 10.1016/S0956-053X(01)00009-5

[6] Sakanakura H. Formation and durability of dithiocarbamic metals in stabilized air pollution control residue from municipal solid waste incineration and melting processes. Environmental Science & Technology. 2007;**41**:1717-1722. DOI: 10.1021/es062077e

[7] Jianguo J, Jun W, Xin X, Wei W, Zhou D, Yan Z. Heavy metal stabilization in municipal solid waste incineration flyash using heavy metal chelating agents. Journal of Hazardous Materials. 2004;**113**:141-146. DOI: 10.1016/j.jhazmat.2004.05.030

[8] Quina MJ, Bordado JCM, Quinta-Ferreira RM. Chemical stabilization of air pollution control residues from municipal solid waste incineration. Journal of Hazardous Materials. 2010;**179**:382-392. DOI: 10.1016/j.jhazmat.2010.03.016

[9] Higuchi S. Technical transitions and future trends at MSW landfill sites. Mater Cycles Waste Manag Res. 2015;**26**:3-11

[10] Hiraoka M, Takeda N. Treatment of sludge containing heavy metals and other harmful substances. Environ Conserv Eng. 1974;**3**(10):33-40

[11] Stegemann JA, Zhou Q. Screening tests for assessing treatability of inorganic industrial wastes by stabilisation/solidification with cement. Journal of Hazardous Materials. 2009;**161**:300-306

[12] Ishimori H, Ishigaki T, Nagamoto K, Yamada K, Yamada M. Leaching amounts of toxic substances from cement- and magnesium-solidified hazardous wastes. Vol. 42. Proceedings of 42th Waste Management Research Symposium, Tokyo, Japan; 2021. pp. 275-277

[13] Fukuda N, Takaoka M, Oshita K, Mizuno T. Stabilizing conditions of metal mercury in mercury sulfurization using a planetary ball mill. Journal of Hazardous Materials. 2014;**276**:433-441

[14] Uenishi R, Kusakabe T, Shiota K, Fujimori T, Oshita K, Takaoka M. Development of solidification technology for sulfurized waste mercury using epoxy

resin. Proceedings of the 10th International Conference on Combustion, Incineration/Pyrolysis, Emission and Climate Change (i-CIPEC) in Bangkok December 18-12, 2018

[15] Ishimori H, Hasegawa R, Ishigaki T. Long-term leaching and volatilization behavior of stabilized and solidified mercury metal waste. Journal of Material Cycles and Waste Management. 2021;**23**: 741-754

[16] Polettini A, Pomi R, Sirini P, Testa F. Properties of Portland cement - Stabilised MSWI fly ashes. Journal of Hazardous Materials. 2001;**88**:123-138. DOI: 10.1016/S0304-3894(01)00292-8

[17] Benassi L, Pasquali M, Zanoletti A, Dalipi R, Borgese L, Depero LE, et al. Chemical stabilization of municipal solid waste incineration fly ash without any commercial chemicals: First pilot-plant scaling up. ACS Sustainable Chemistry & Engineering. 2016;**4**:5561-5569. DOI: 10.1021/acssuschemeng.6b01294

[18] Dermatas D, Xu X, Cao X, Shen G, Menounou N, Arienti P, et al. An evaluation of pozzolanic lead immobilization mechanisms in firing range soils. Proc Int Conf Stabilisation/Solidification Treat Remediat. 2005:45-56. DOI: 10.1201/9781439833933.ch8

[19] Gougar MLD, Scheetz BE, Roy DM. Ettringite and C-S-H Portland cement phases for waste ion immobilization: A review. Waste Management. 1996;**16**: 295-303. DOI: 10.1016/S0956-053X(96)00072-4

[20] Chen QY, Tyrer M, Hills CD, Yang XM, Carey P. Immobilisation of heavy metal in cement-based solidification/stabilisation: A review. Waste Management. 2009;**29**:390-403. DOI: 10.1016/j.wasman.2008.01.019

[21] Wang L, Jin Y, Nie Y. Investigation of accelerated and natural carbonation of MSWI fly ash with a high content of Ca. Journal of Hazardous Materials. 2010; **174**:334-343. DOI: 10.1016/j.jhazmat.2009.09.055

[22] Karlfeldt Fedje K, Ekberg C, Skarnemark G, Steenari BM. Removal of hazardous metals from MSW fly ash-An evaluation of ash leaching methods. Journal of Hazardous Materials. 2010; **173**:310-317. DOI: 10.1016/j.jhazmat.2009.08.094

[23] Shim Y, Rhee S, Lee W. Comparison of leaching characteristics of heavy metals from bottom and fly ashes in Korea and Japan. Waste Management. 2005;**25**:473-480. DOI: 10.1016/j.wasman.2005.03.002

[24] Alexander GB, Heston WM, Iler RK. The solubility of amorphous silica in water. The Journal of Physical Chemistry. 1954;**58**:453-455. DOI: 10.1021/j150516a002

[25] Kitamura H, Ueshima M, Back S, Sutthasil N, Sakanakura H, Ishigaki T, et al. Impact of diatomite addition on lead immobilization in air pollution control residues from a municipal solid waste incinerator. Environmental Science and Pollution Research. 2021. DOI: 10.1007/s11356-021-17349-x

[26] Kastis D, Kakali G, Tsivilis S, Stamatakis MG. Properties and hydration of blended cements with calcareous diatomite. Cement and Concrete Research. 2006;**36**: 1821-1826. DOI: 10.1016/j.cemconres.2006.05.005

[27] Ergün A. Effects of the usage of diatomite and waste marble powder as partial replacement of cement on the mechanical properties of concrete. Construction and Building Materials.

2011;**25**:806-812. DOI: 10.1016/j.conbuildmat.2010.07.002

[28] Xu S, Wang J, Jiang Q, Zhang S. Study of natural hydraulic lime-based mortars prepared with masonry waste powder as aggregate and diatomite/fly ash as mineral admixtures. Journal of Cleaner Production. 2016;**119**:118-127. DOI: 10.1016/j.jclepro.2016.01.069

[29] Li J, Zhang W, Li C, Monteiro PJM. Green concrete containing diatomaceous earth and limestone: Workability, mechanical properties, and life-cycle assessment. Journal of Cleaner Production. 2019;**223**:662-679. DOI: 10.1016/j.jclepro.2019.03.077

[30] Escalante-García JI, Sharp JH. Effect of temperature on the hydration of the main clinker phases in Portland cements: Part i, neat cements. Cement and Concrete Research. 1998;**28**:1245-1257. DOI: 10.1016/S0008-8846(98)00115-X

[31] Greenberg A. Reaction between silica and calcium hydroxide solutions. I. Kinetics in the temperature range to 30 to 85°C. The Journal of Physical Chemistry. 1961;**65**:12-16. DOI: 10.1021/j100819a005

[32] Assi A, Bilo F, Zanoletti A, Ponti J, Valsesia A, La Spina R, et al. Zero-waste approach in municipal solid waste incineration: Reuse of bottom ash to stabilize fly ash. Journal of Cleaner Production. 2020;**245**:118779. DOI: 10.1016/j.jclepro.2019.118779

[33] Sun X, Li J, Zhao X, Zhu B, Zhang G. A Review on the management of municipal solid waste fly ash in American. Procedia Environmental Sciences. 2016;**31**:535-540. DOI: 10.1016/j.proenv.2016.02.079

[34] Erses AS, Onay TT. In situ heavy metal attenuation in landfills under methanogenic conditions. Journal of Hazardous Materials. 2003;**99**:159-175. DOI: 10.1016/S0304-3894(02)00354-0

[35] Meima JA, Comans RNJ. Geochemical modeling of weathering reactions in municipal solid waste incinerator bottom ash. Environmental Science & Technology. 1997;**31**: 1269-1276. DOI: 10.1021/es9603158

[36] Piantone P, Bodenan F, Chatelet-Snidaro L. Mineralogical study of secondary mineral phases from weathered MSWI bottom ash: Implications for the modelling and trapping of heavy metals. Applied Geochemistry. 2004;**19**:1891-1904. DOI: 10.1016/j.apgeochem.2004.05.006

[37] Ferrari S, Belevi H, Baccini P. Chemical speciation of carbon in municipal solid waste incinerator residues. Waste Management. 2002;**22**: 303-314. DOI: 10.1016/S0956-053X(01)00049-6

[38] Furukawa T, Takahashi Y. Oxalate metal complexes in aerosol particles: Implications for the hygroscopicity of oxalate-containing particles. Atmospheric Chemistry and Physics. 2011;**11**:4289-4301. DOI: 10.5194/acp-11-4289-2011

[39] Sugiura Y, Suyama T, Tachi Y. Development of JAEA sorption database (JAEA-SDB): Update of sorption/QA Data in FY2019. JAEA-Data/Code 2019–022 2020; Available from: https://migrationdb.jaea.go.jp/sdb_j2/sdb_pre_j.html

[40] Agrafioti E, Kalderis D, Diamadopoulos E. Arsenic and chromium removal from water using biochars derived from rice husk, organic solid wastes and sewage sludge. Journal of Environmental Management. 2014; **133**:309-314

[41] Allison JD, Allison TL. Partition coefficients for metals in surface water, soil, and waste. Washington, US. EPA/600/R-05/074.2005; 2005

[42] Duc TA, Loi VD, Thao TT. Partition of heavy metals in a tropical river system impacted by municipal waste. Environ Monitor Assess. 2013;**185**:1907-1925

[43] Feng C, Guo X, Yin S, Tian C, Li Y, Shen Z. Heavy metal partitioning of suspended particulate matter-water and sediment-water in the Yangtze Estuary. Chemosphere. 2017;**185**:717-725

[44] Janssen RPT, Peijnenburg WJGM, Posthuma L, Van Den Hoop MAGT. Equilibrium partitioning of heavy metals in Dutch field soils. I. Relationship between metal partition coefficients and soil characteristics. Environmental Toxicology and Chemistry. 1997;**16**: 2470-2478

[45] Keshavarzi B, Moore F, Sharifi R. Evaluation of dispersion and chemical partitioning patterns of heavy metals in the Sar Cheshmeh porphyry copper deposit: geochemical data from mine waste, water and stream sediments. Int J Environ Studies. 2013;**70**:73-93

[46] Sheppard S, Long J, Sanipelli B. Solid/liquid partition coefficients (Kd) for selected soils and sediments at Forsmark and Laxemar-Simpevarp. Stockholm, Sweden: Geological Survey of Sweden. SKB Rapport R-09-27; 2009

[47] Yun-Zeng C, Hao Y, Zhen-Ke Z, Ming-Zhou Q, Feng J, Jun-Jie LU. Application of equilibrium partitioning approach to the derivation of sediment quality guidelines for metals in Dianchi lake. Pedosphere. 2007;**17**:284-294

[48] Krupka KM, Kaplan DI, Whelan G, Serne RJ, Mattigod SV. Understanding variation in partition coefficient, Kd, values. Washington, DC, USA. EPA/402/R-99/004B; 1999

[49] Sheppard S, Sohlenius G, Omberg LG, Borgiel M, Grolander S, Norden S. Solid/liquid partition coefficients (Kd) and plant/soil concentration ratios (CR) for selected soils, tills and sediments at Forsmark. Geological Survey of Sweden. Stockholm, Sweden. SKB Rapport R-11-24; 2011

[50] Thibault DH, Sheppard MI, Smith PA. A critical compilation and review of default soil solid/liquid partition coefficients, Kd, for use in environmental assessments. Pinawa, Canada: Atomic Energy of Canada Limited. AECL-10125; 1990

[51] Kuhlmeier PD. Partitioning of arsenic species in fine-grained soils. J Air & Waste Manage Assoc. 1997;**47**: 481-490

[52] Tabelin CB, Igarashi T, Arima T, Sato D, Tatsuhara T, Tamoto S. Characterization and evaluation of arsenic and boron adsorption onto natural geologic materials, and their application in the disposal of excavated altered rock. Geoderma. 2014;**213**: 163-172

[53] Yang X, Hou Q, Yang Z, Zhang X, Hou Y. Solid-solution partitioning of arsenic (As) in the paddy soil profiles in Chengdu Plain. Southwest China. Geoscience Frontiers. 2012;**3**(6):901-909

[54] Hassan SM, Garrison AW. Distribution of chromium species between soil and porewater. Chemical Speciation & Bioavailability. 1996;**8**: 85-103

[55] Sheppard SC. Robust prediction of Kd from soil properties for

environmental assessment. Human and Ecological Risk Assessment. 2011;**17**: 263-279

[56] Kumpiene J, Lagerkvist A, Maurice C. Stabilization of As, Cr, Cu, Pb and Zn in soil using amendments - A review. Waste Management. 2008;**28**: 215-225

[57] Wada S. Dynamics of heavy metals in soils. Chikyu Kankyo. 2010;**15**:15-21

[58] Shimada N. The essence of problems on groundwater and soil pollutions caused by naturally occurring heavy metals and harmful elements: lead. Technical report in OYO Corporation. 2013:1-27

[59] Gregory DD, Large RR, Halpin JA, Baturina EL, Lyons TW, Wu S, et al. Trace element content of sedimentary pyrite in black shales. Economic Geology. 2015;**110**:1389-1410

[60] Kim D, Cai Z, Sorial GA. Determination of gas phase adsorption isotherms - A simple constant volume method. Chemosphere. 2006;**64**: 1362-1368

[61] Wang Y, Raihala TS, Jackman AP, StJohn R. Use of Tedlar bags in VOC testing and storage: evidence of significant VOC losses. Environmental Science & Technology. 1996;**30**: 3115-3117

[62] Ishimori H, Suzuki T. Sakanakura H, Ishigaki T. Establishing soil adsorption testing methods for gaseous mercury and evaluating the distribution coefficients of silica sand, decomposed granite soil, mordenite, and calcium bentonite. Soils and Foundations 2020;**60**:496-504

[63] Ishimori H, Hasegawa R, Endo K, Sakanakura H, Ishigaki T. Numerical simulations of leaching and volatilization behaviors from stabilized and solidified mercury metal waste in landfill. Global Environ Res. 2020;**24**(1):11-18

[64] Tojo Y, Ikeda T, Matsuo T, Matsuto T. The effect of inside circumstance of the hazardous waste landfill on the leaching behavior of harmful heavy metals. Int J Geomate. 2020;**18**(66):129-134

[65] Ishimori H, Endo K, Ishigaki T, Yamada K, Yamada M. Multiphysics of seismic analysis and reactive chemical transport analysis for long-term environmental safety evaluation of RC-built hazardous waste landfill. Jap Geotech Soc Special Publication. 2021;**9**: 136-141

[66] Edvardsen C. Water permeability and autogenous healing of cracks in concrete. ACI Materials Journal. 1999;**96**: 448-454

[67] Japanese Society of Civil Engineering. Standard specifications for concrete structures. Tokyo, Japan: JSCE; 2018

[68] European Committee for Standardization. Eurocode 2: Design of concrete structures - Part1–1: General rules and rules for buildings. Brussels, Belgium. BSN1992-1-1; 2004

[69] American Concrete Institute. Building code requirements for structural concrete and commentary. Farmington Hills, USA. ACI318–05 and ACI318R-05; 2005

Chapter 9

Technical Cooperation for Enhancing Infectious Healthcare Waste Management

Mitsuo Yoshida

Abstract

Appropriate healthcare waste (HCW) management is crucial for preventing the spread of infectious diseases and ensuring public health. However, in many economically developing countries, HCW is often insufficiently segregated at sources and the hazardous infectious components are mixed with municipal solid waste, and directly disposed without any treatment, which poses great risk factors for healthcare institutes, waste management service providers, and the public. According to statistical cross-country analysis, the amount of HCW generation is expected to increase sharply in near future in developing countries. What can we do to improve the status of HCW management in developing countries and prevent the spread of infection? It is necessary to establish an effective HCW management system and strengthen its implementation capacity, especially in developing countries. When conducting international technical assistance for them to support the establishment of the HCW management system and capacity development, it is required to set the targets for technical assistance through conducting an assessment survey, analyzing problems, evaluating risks, supporting to formulate management plans, and provision of equipment. A diagnosis method for existing capacity and challenges is proposed for planning technical assistance. Experiences of technical assistance on HCW management in Palestine are presented as a case study.

Keywords: healthcare waste management, developing countries, diagnosis of healthcare waste management system, technical assistance

1. Introduction

Healthcare waste (HCW), which is defined as all the waste generated within healthcare institutes, research institutes, laboratories related to medical procedures, and healthcare activities in the home [1], contains infectious hazardous waste components and must be properly sorted, collected, and treated to prevent infection [2, 3].

Over the last decades, the need for safety management of HCW has significantly increased due to the rapid population growth and increase in medical institutes, without which potential risks are very high to human health and the environment. Approximately five million people were reported to die every year due to HCW-related diseases [4]. The risk comes from accidental injuries during the handling of infectious waste components of HCW, which can cause diseases like hepatitis B, hepatitis C, and HIV infection [5]. Moreover, numerous other diseases can be transmitted by contact with infectious HCW.

The importance of HCW management is once again drawing attention in today's COVID-19 pandemic [6]. As the virus stays longer on plastic, metal, and cardboard materials in HCW generated from the treatment of COVID-19 infected patients can be one of the potential routes for transmission of infection [7, 8].

However, in economically developing countries, which was defined by United Nations [9], HCW management systems are often unestablished or not fully functional, where infectious HCW that are not properly segregated and/or treated turn into new sources of infection and the waste streams become as a path to spread the infection. According to WHO, just over half (58%) of the sampled facilities from 24 countries (as of 2015) had adequate systems in place for the safe disposal of HCW generated [10]. Many papers have been reported that the implementation system and capacity of HCW management are often inadequate [4, 5, 11–18].

What can we do to improve the status of HCW management in developing countries and prevent the spread of infection? Especially in low-income developing countries where funds are insufficient, capacities at organizational, institutional, and societal levels are weak, some technical assistance for enhancing HCW management will be required through international cooperation [10]. In order to make effective use of the limited resources of donor agencies and provide appropriate technical assistance, it is essential to understand the current situation, evaluate risks, diagnose existing systems, and provide necessary technical supports.

The main theme of this chapter is to consider the challenges of international technical assistance for improving the management of hazardous infectious HCW in developing countries. The author first analyzes the trends in medical waste generation around the world and predicts the needs for medical waste management in developing countries in the near future. Next, we will examine the problems of HCW management in developing countries and the risks arising from them, and review the solutions to the problems, the system design, and an effective HCW management plan for avoiding the risks. It also proposes diagnostic methods and necessary supports for implementing effective international technical assistance. Finally, a case study of the international technical assistance on HCW management in Palestine is described.

2. HCW generation in developing countries

2.1 Definition of HCW

First, the term HCW used in this chapter will be clarified in detail based on the WHO definition [1]. HCW is a broad concept and can be classified into 2 main categories, hazardous and non-hazardous HCW. According to the definition by WHO [1], the hazardous HCW can be subdivided into 6 subcategories, as follows (**Figure 1**).

1. Sharps waste: It is defined as used or unused sharps, such as hypodermic, intravenous, or other needles; auto-disable syringes; syringes with attached needles; infusion sets; scalpels; pipettes; knives; blades; broken glass.

2. Infectious waste (narrow sense): It is suspected to contain pathogens and that poses a risk of disease transmission, for example, waste contaminated with

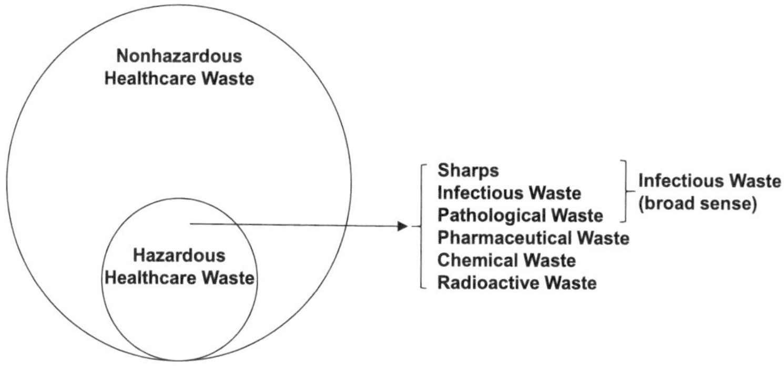

Figure 1.
Classification of healthcare waste (HCW) based on the definition of [1]. About 15% of HCW generated is hazardous while the remaining 85% is the same as general municipal waste (MSW).

blood and other body fluids; dressings, bandages, swabs, gloves, masks, gowns, drapes and other material contaminated with blood or other body fluids; laboratory cultures and microbiological stocks; and waste including excreta.

3. Pathological waste: This is consist of human tissues, organs, or fluids; body parts; fetuses; unused blood products.

4. Pharmaceutical/cytotoxic wastes: These are any waste that contains medical drugs that are expired, unused, or no longer needed.

5. Chemical waste: It is regulated as hazardous waste if it exhibits one of four characteristics: ignitability, corrosive, reactivity, or ability to produce toxic leachate in a landfill.

6. Radioactive materials: This has proven to be valuable tools in medicine, while eventually becoming low-level radioactive waste.

Among these 6 subcategories, (1) sharps, (2) infectious, and (3) pathological wastes are at risk of disease transmission, which are often collectively referred to as 'infectious waste' in the broad sense of the term. In general, around 10% of HCW is infectious waste in the broad sense, 85% of HCW is non-hazardous general waste, and 5% are the other hazardous ones ((4) pharmaceutical, (5) chemical, and (6) radioactive) those must be distinguished from other HCW and properly treated and disposed based on the national regulation and standard.

2.2 Increase of HCW generation according to economic growth

The characteristics of HCW generation depend on economic and social conditions, public health conditions, healthcare service systems, and solid waste management systems in each country.

According to the World Bank [19], the global average of HCW generation is 0.25 kg/capita/day, which is a very small part of the total special waste generated. **Figure 2** shows the relationship between the level of economic growth (GDP (USD)/capita) and the HCW generation rate (kg/capita/year), using the data given by the World Bank [19] and JICA.

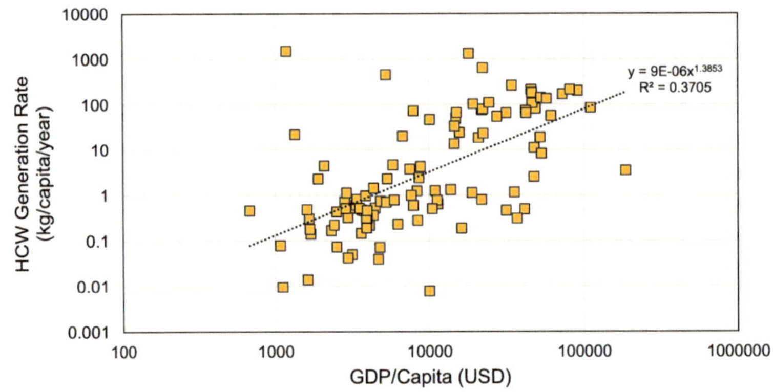

Figure 2.
Correlation between economic growth (GDP/capita) and HCW generation rate (kg/capita/year). Each plot indicates country averaged data (2011–2017), and the dashed line is the trend. Both horizontal and vertical axes are on a logarithmic scale.

HCW generation data in 105 countries/regions are available and are plotted in the diagram as a cross-country analysis (**Figure 2**). As is clear from this diagram, the level of economic growth (GDP/capita) and HCW generation rate (kg/capita/year) show a weak positive power correlation (r^2 = 0.3705). In relatively low income countries (GDP/capita <10,000 USD), HCW generations are often less than 1.0 kg/capita/year.

A similar correlation can be observed for the relationship between the total amount of municipal solid waste (MSW) generated in each country/region and the total amount of HCW generated, as shown in **Figure 3**, in which the data set used was the same as **Figure 2**.

The total amount of MSW generated and the total amount of HCW generated are in a positive power correlation (r^2 = 0.5277), which is stronger than

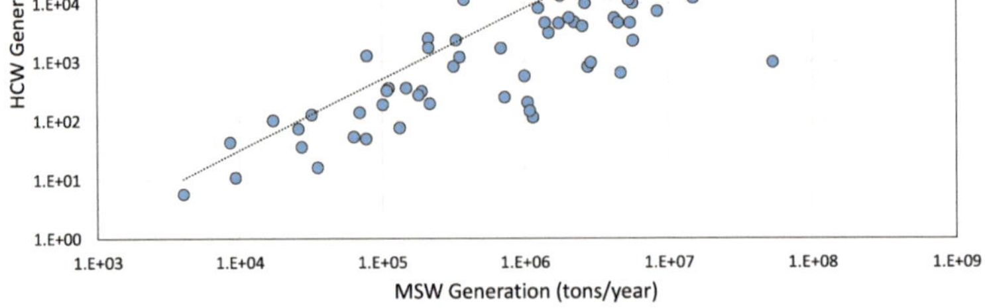

Figure 3.
Correlation between HCW and municipal waste (MSW) generations. Each plot indicates country averaged data (tons/year; 2011–2017), and the dashed line is the trend. Both horizontal and vertical axes are on logarithmic scale.

the above-mentioned correlation with economic growth. The diagram indicates that the generation status of MSW has a strong influence on the generation status of HCW.

It has been reported that the amount of MSW generated increases year by year due to factors such as economic growth, diversification of life, urbanization, and population growth, in which the rate of increase in developing countries is higher than that in developed countries [19]. This indicates that enhancing HCW management capacity is a strong need in developing countries.

2.3 HCW and waste management capacity

Recognizing the necessity of HCW management by government authorities means understanding its risks and considering the need for proper collection, treatment, and disposal. In that sense, it can be said that the increase in official reporting of the amount of HCW is related to the improvement of authorities' concerns on HCW management.

Figure 4 shows the MSW collection service coverage rate (%; based on the target population or the total amount of waste generated) and the amount of HCW generation (kg/capita/year) that is officially recognized by each government authority. The MSW service coverage rate can be used as an index for the quality of MSW management services.

As is clear from this figure, the HCW generation rate is generally very low, if the MSW service coverage rate is less than 83%. In other words, it shows that government authority and given administration system cannot properly respond to HCW generation unless the services for MSW management are in place to some extent. Conversely, in countries/regions where the service coverage rate is more than 83%, the necessity of enhancing HCW management is emphasized as the high priority issue for relevant authorities.

As we have seen earlier, the amount of HCW generated is closely related to the degree of economic development and also to the state of the MSW management services. It shows that socially recognizing HCW as hazardous waste and implementing necessary treatment and disposal will gradually develop in accordance with the enhancement of the capacity of MSW management service as well as economic development.

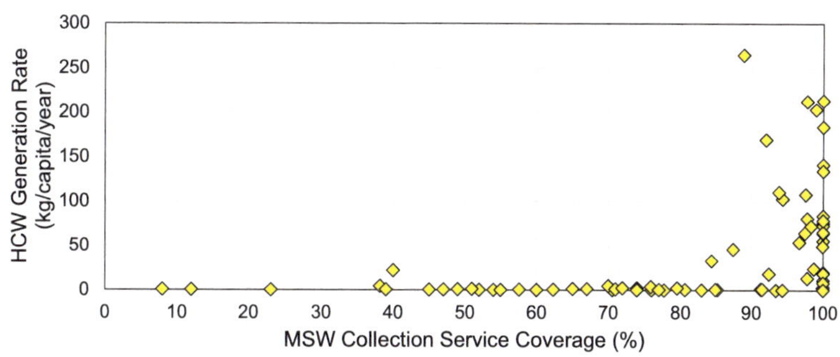

Figure 4.
Correlation between MSW collection service coverage (% in population or total waste basis) and HCW generation rate (kg/capita/year).

3. Risk, problem, and capacity in HCW management

3.1 General concept of HCW management

The purposes of solid waste management (SWM) are similar, whether addressing hazardous, infectious, or even general municipal waste; three themes are prominent; management, treatment, and waste minimization [20]. The management of HCW requires analysis and active control from generation to final disposal. Hazardous or infectious HCW should be appropriately treated before disposal to eliminate its hazard risks. Waste minimization or reduction is undoubtedly the most desirable goal of solid waste management, which is the same in HCW management, where 3Rs (Reduce, Reuse, Recycle) are key approaches after segregation at source and appropriate treatment of hazardous infectious components (**Figure 5**).

3.2 Risks caused by HCW

It is crucial that decision-makers and administrators have a complete understanding of the risks of hazardous HCW since they are responsible for setting the HCW management system with a safe workplace and preventing environmental pollution. Inappropriate HCW management poses five major occupational, health, and environmental risks; A, B, C, D, and E, as shown in **Figure 6**.

A. Risks of infection by HCW within the healthcare institutes

The most common and most investigated cause of the microbiological risks associated with HCW are injuries due to needles of sharps waste [21]. For example, according to the results of a questionnaire survey of HCW workers engaged in Palestine, 32% in-hospital workers and 27% SWM workers experienced some kind of infectious waste accident when the HCW management system had not been established and the staff training had not been given [22]. According to the CDC guideline [23], a leak-resistant biohazard bag is usually adequate for containment of infectious wastes, and puncture-resistant

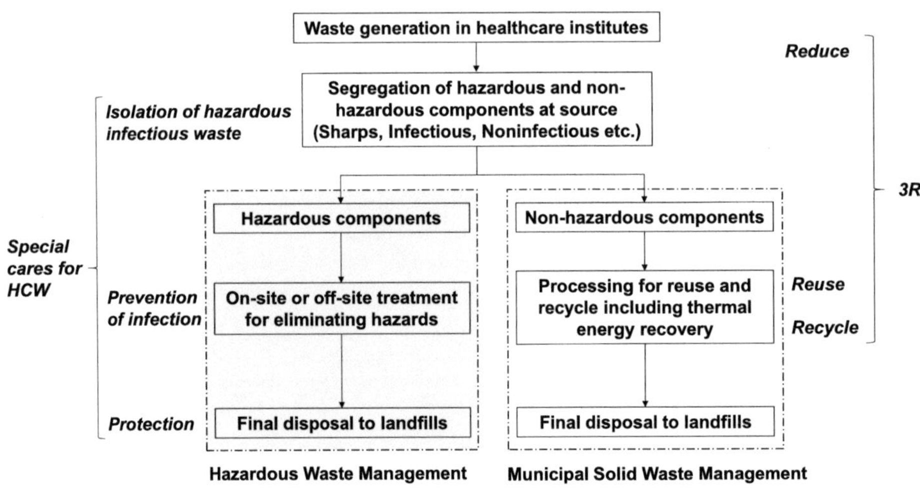

Figure 5.
HCW management and the concept of 3Rs (Reduce, Reuse, recycle).

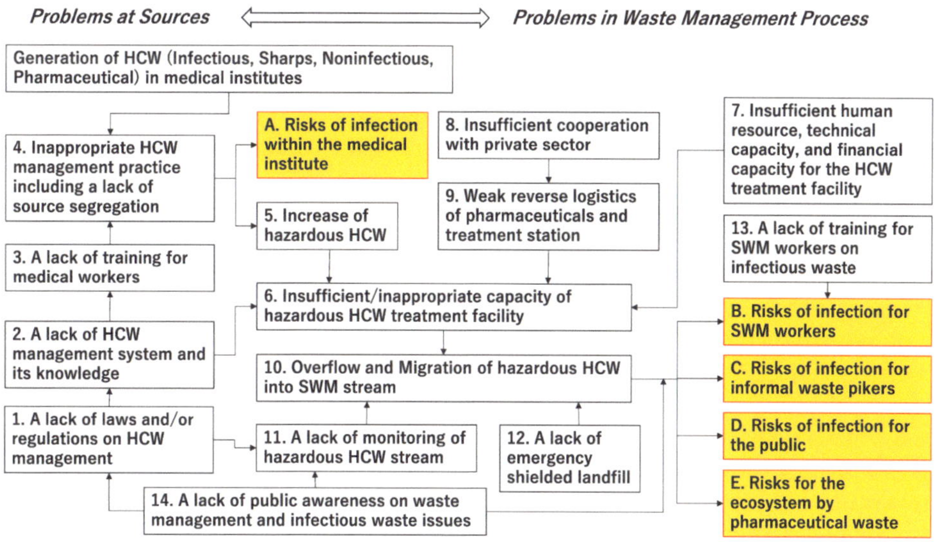

Figure 6.
Causal linkage of problems and risks in hazardous HCW management often observed in developing countries. White boxes show a series of problems in HCW management and the colored boxes indicate the risks.

containers located at the point of use are set as containment for discarded tubes with small amounts of blood, scalpel blades, needles & syringes, and other sharps.

B. Occupational risks for waste management workers

The impact of hazardous HCW in developing countries is very likely to pose a great occupational risk to general SWM workers and the public outside healthcare institutes due to inadequate practices of SWM and personal protection for workers themselves. In addition, the hazards posed by HCW may be more significant due to the limited availability of immunization against infectious diseases. The distribution of personal protective equipment (PPE) such as gloves, goggles, facemask, and disinfectant, is effective to prevent accidental infections together with periodical guidance and training.

C. Risks of infection for informal waste pickers

In countries where the HCW management system is not established nor functional, HCW is directly dumped at dumpsites without any treatment. These include infectious waste and sharps, and when waste pickers collect recyclables from the sites, they can cause injury and eventually infection. The existence of informal waste pickers is due to socio-economic problems and is not directly related to the HCW management issue. However, recognizing the existence of such risks, even if the only way is direct disposal of infectious HCW, the dumping site should be off-limits or immediately covered with soil for avoiding direct exposure.

D. Risks of infection to the public

Analysis of the microbiological content of MSW and HCW has shown similar concentrations of microorganisms in both types of wastes. According to the microbiological study of HCW and MSW, 2% of blood-stained waste was positive for hepatitis viruses, and poliovirus and echovirus were recovered from soiled diapers in MSW [24]. Some infectious waste can stay infectious for many years if disposed without being sterilized. For example, anthrax-infected cattle contain spores that are known for many decades in dry soil [20, 25]. Therefore, dumping site management is required if infectious HCW must be directly disposed without any treatment.

E. Risks of hazardous substances for the ecosystem

If pharmaceutical or chemical pollutants are released into the environment, they can easily diffuse through groundwater, surface water, and eventually leach into drinking water. Pharmaceuticals are discarded and renewed in healthcare facilities when they expire. In time of conflicts or natural disasters, large quantities of pharmaceuticals are often donated as a part of humanitarian assistance [26]. However, such donated pharmaceuticals are sometimes stocked and often mismanaged when the pharmaceutical management system is not well functioned. Disposal of these unwanted or expired drugs may disturb the ecosystem. WHO guidelines [1, 27] recommended pharmaceutical waste to reverse distributors.

Problems	Description	Challenges
1. A lack of laws and/ or regulations for HCW management	Due to the unclear definition and management responsibilities of HCW, some HCWs are not processed or not properly managed. Duplication and fragmentation impede system efficiency	Establishment of the legal system on HCW. Definition of HCW and clarification of management responsibilities in line with the current situation. Establishment of HCW treatment standards
2. A lack of HCW management system and its knowledge	HCW management system has not yet been established, where no HCW management plan and unclear implementation body are determined. Plan-Do-Check-Act (PDCA) management cycle is not developed. Inefficiencies occur due to unplanned waste treatment activities	Establishment of HCW management system with implementation body and HCW management plan
3. A lack of training for medical workers	Due to inadequate training for medical workers, they have insufficient knowledge on the dangers of hazardous HCW and the precautions to be taken when handling HCW. As a result, an infection accident occurs within the medical institute	Create manuals, textbooks, and teaching materials for staff training. Train staff training instructors. Organize regular staff training
4. Inappropriate HCW management practice including a lack of source segregation	Insufficient source separation increases infectious HCW and increases the loads for the treatment system. This is because if non-infectious waste is discharged without being separated from infectious waste, all becomes infectious waste. This results in the overloading of hazardous HCW to the existing treatment system, where the incoming waste amount to the treatment system exceeds its planned capacity	Training for staff in the medical institute, preparation of manual and posters for source separation. Inspection system for the situation of source separation and container management
5. Increase of hazardous HCW		Encouraging the source separation practice
6. Insufficient/ inappropriate capacity of hazardous HCW treatment facility		Enhancing the treatment system

Problems	Description	Challenges
7. Insufficient human resource, technical capacity, and financial capacity for the hazardous HCW treatment facility	There is a lack of human resources, technology, and financial base to establish a proper treatment system of hazardous HCW	Provision of equipment technology transfer, and training Financial support for establishing the treatment system
8. Insufficient cooperation between public, medical, and private sectors about hazardous HCW issue	The best available technology is required for proper treatment of hazardous HCW in given country conditions, and the private sector plays a large role in introducing it. It is expected that the suppliers of pharmaceuticals handle unused/wasted them, but it cannot be dealt without sound cooperation between the public, medical, and private sectors	Networking between public, medical, and private sectors. Defining a rule for treatment of pharmaceutical waste in HCW. Establishing a reverse logistic system for pharmaceuticals. Establishing a treatment facility for pharmaceuticals and chemicals
9. Weak reverse logistics of pharmaceuticals and/or treatment		
10. Overflow and migration of hazardous HCW into municipal SWM stream	As a result of the increase in the amount of hazardous HCW results in the overflow of HCW containing hazardous components, which poses a risk of spreading infection and other negative impacts through the waste stream	Monitoring of the treatment system
11. A lack of monitoring of hazardous HCW stream	Despite the inclusion of hazardous waste, general waste treatment and disposal is carried out without this in mind. This deteriorates the occupational health and safety conditions of SWM workers	Monitoring of storing, transportation, and final disposal of hazardous HCW, using a manifest system
12. A lack of emergency shielded landfill	If the necessary sterilization or treatments are not possible, the untreated hazardous HCW will be directly landfilled as an emergency measure. In that case, the disposal site must be shielded, otherwise, infection and/or contamination can spread throughout the environment	Constructing shielded landfill Landfill operation and management
13. A lack of training on infectious HCW for SWM workers	As mentioned above, there are various possibilities that hazardous HCW will migrate into the general waste stream, and especially in the COVID-19 pandemic, the general waste management flow itself can also be a path of infection. It is necessary to train workers engaged in SWM service to prevent infection and to use PPE	Training for SWM workers about health and safety conditions. Distribution of personal protective equipment (PPE) for SWM workers
14. A lack of awareness on waste management and hazardous HCW issues	Behind all the above issues lies the issue of awareness of the HCW issue. This includes not only the general public but also workers and decision-makers at various levels of society	Public awareness-raising on HCW and its risks Awareness-raising for decision-makers

Table 1.
Problems recognized in HCW management and challenges for solving them.

3.3 Problems and challenges in HCW management

The above-mentioned five risks are caused by 14 problems in HCW management, and the relationship between the risks and problems is depicted as a causal linkage diagram as shown in **Figure 6**. The description of each problem and the challenges for enhancing HCW management, which indicate the goal of technical assistance, are summarized in **Table 1**. If HCW waste is not properly segregated at the source (Problems No. 1–4), it creates many risks associated with various technical and management factors in the HCW management process (No. 5–14).

4. Establishing effective HCW management

4.1 Institutional development for HCW management

In designing a system for HCW management, it is important to consider three layers institutions levels; namely, global, national, and local levels. The institution at the global level is given by the internationally-accepted guidelines published by WHO [1] and other international organizations.

In any country, a national policy is the first step in creating a successful and sustainable HCW management system. The policy should be the blueprint to drive decision-making at a political level, for the allocation of resources, and mobilize government efforts to create the conditions to implement an HCW management system [28]. Based on the international guidelines, national-level legal systems and institutions for conducting proper HCW management will be formulated according to the given conditions of the country. Specific and comprehensive legislation and policy documents on HCW management with a clear designation of responsibilities to various stakeholders are required [18].

The following five basic principles are important in formulating an effective HCW management system, which was originally specified by the Global Healthcare Waste Project conducted by the United Nations Development Programme (UNDP) in cooperation with the Global Environment Facility (GEF) and WHO [29]:

i. Polluter pays principle: All waste generators are legally and financially responsible for safe handling of waste and environmentally sound disposal of waste.

ii. Precautionary principle: In order to protect the environment, the precautionary approach [30] shall be applied according to their capabilities.

iii. Duty of care principle: Stipulates that any person handling or managing hazardous HCW is ethically responsible for applying the utmost care.

iv. Proximity principle: Treatment and disposal of hazardous HCW take place as near as possible to the generation point for minimizing potential risks during transportation.

v. Prior informed consent principle: Prior informed consent is required for the siting and operation of HCW treatment facilities.

4.2 Planning HCW management

There are basically two types of HCW management; a national level HCW management and a local level (individual healthcare institute or service provider) HCW management. In some cases, regional (provincial, prefectural) level HCW management is set between the national and local levels.

4.2.1 National/regional HCW management plan and strategy

The purpose of planning a national/regional HCW management is to improve HCW management at the national and regional (e.g., provincial) levels, where strong political commitment is required. In the planning process, it is required to involve relevant ministries and professional organizations including academics in the HCW management field.

The goals of the national/regional HCW management plan are: to declare the government's intentions to improve HCW management, to define overall national/regional strategies and plan for improving HCW management, to specify activities and timeline for implementation, and to define the roles and responsibilities of authorities concerned & other stakeholders.

1. Assessment study

The first step for formulating a national HCW management plan is to conduct a national assessment study on HCW management, where the following four points have to be clarified: (i) an inventory of existing healthcare institutes (waste generators) and HCW treatment facilities; (ii) analysis of existing legislation, regulations, and rules; (iii) existing HCW stream and its management practices if any; and (iv) implementation agency and human resource on HCW management.

The inventory of HCW generation sources and GIS (geographical information system) map are crucial for planning HCW management, which covers all healthcare institutes including hospitals, clinics, and primary healthcare (PHC) institutions. A regression model will be applied to estimate the amount of HCW generated by them using the outpatient, inpatient, and bed numbers.

2. Planning

Specific objectives toward developing a national HCW management plan should include the following five key objectives: (1) to understand the present situation and setting the purpose of the plan, (2) to develop the legal and regulatory framework, (3) to develop financial investments and resources for HCW management, (4) to develop capacity building program, and (5) to set up a monitoring plan. The expected general contents of the national HCW management plan are as shown in **Table 2**.

4.2.2 Local HCW management plan

When planning a local HCW management, the first thing that must be done is to clarify the executing agency based on the legal system, and that agency will make the plan. The local HCW management plan is created by each healthcare institute and/or service provider based on the above-mentioned national/regional plan. It is required to be specific and practical depending on given local conditions.

Specific objectives toward developing a local HCW management plan needs to include the following six key components based on the direction of the national/regional HCW management plan: (1) designing HCW management system, (2) segregation of HCW at source, (3) HCW handling, storage, and transport, (4) treatment technologies, (5) waste disposal, (6) staff training, and (7) monitoring:

Categories	Contents
(1) To understand the background and setting the purpose of the plan	• Summary of the results of Assessment Study (see 4.2.1 (1)) • Issue and challenges on HCW management • Setting the purpose and goals of the HCW management plan
(2) To develop the legal and regulatory framework	• Present state of HCW management, issues and challenges • Legal framework, laws, regulations, standards on HCW • Establishment of implementing agency • Coordination with other laws such as public health, SWM, and environmental protection • Guideline(s) for proper HCW management • Method of enforcement • Licensing system for healthcare institutes, treatment facilities, and service providers
(3) To develop financial investments and resources for HCW management	• Specific budget lines, subsidies, and funds for developing HCW management • Cost recovery mechanisms to sustain HCW management • Public-private partnership • Investment plan
(4) To plan capacity building program	• Training programs including training of trainers (ToT) • National awareness-raising campaigns for target groups • Partnership with healthcare-related professional societies, educational institutes, and universities
(5) To set up a monitoring plan and information strategy	• Define indicators of achievement or performance • Monitoring and inspection of healthcare institutes and service provider • HCW tracking (manifest) system • Database system and information disclosure

Table 2.
General contents of the national HCW management plan.

1. Designing HCW management system

Regarding the treatment and/or sterilization of infectious HCW, there are basically two types of HCW management systems; distributed (on-site) and centralized (off-site) systems. The distributed one is a system in which a healthcare institute has its own (relatively small-scale) treatment facility and processes infectious HCW by itself. On the other hand, the centralized system is a system in which a private or public service provider collects infectious HCW and transports it from each healthcare institute based on a contract and centrally processes it in the service provider-owned treatment facility.

The advantages of the distributed system are: complete control of infectious HCW by the generator, mitigating the risk of exposure during waste collection & transportation, and reducing unknown risks in the treatment by a service provider. However the distributed system has the following disadvantages: the healthcare institute has to get a relatively high financial burden for a treatment facility, and also become responsible for meeting all regulatory requirements on infectious HCW treatment, which needs additional resources.

On the other hand, the advantages of a centralized system are minimization of cost and responsibility for the treatment of infectious HCW. Each healthcare institute can concentrate only on source separation and appropriate waste discharging. The HCW generators, in particular small-scale healthcare institutes, benefit from the quality of service and the economies of scale. The disadvantages are indirect control of infectious HCW management due to outsourcing.

In the case of a centralized system, two management plans, on-site and off-site plans, are required, and coordination and cooperation between the two actors are indispensable.

These distributed and centralized systems are often combined in a country/region to act as a hybrid system for actual HCW management (**Figure 7**). As shown in the **Figure 7**, the generated HCW is first separated into non-infectious HCW and infectious HCW (sharps and infectious) at the source. Non-infectious HCW is treated in the MSW management flow, while infectious HCW is sent for on-site or off-site treatment. If there is no treatment facility or in the case of the treatment capacity is insufficient, an emergency controlled cell is installed at the landfill site for direct disposal as an emergency measure.

2. Segregation at source

Segregation at source (source separation) is one of the most important steps to successfully manage HCW. As shown in **Figure 1**, only about 15% of the HCW is hazardous, treatment and disposal costs could be greatly reduced if proper segregation were performed. Segregating hazardous from nonhazardous waste reduces also greatly the risks of infecting SWM workers.

Segregation consists in separating the different waste streams based on the hazardous properties of the waste, the type of treatment and disposal practices that are applied. A recommended way of identifying HCW categories is by sorting the waste into color-coded and well-labeled bags or containers.

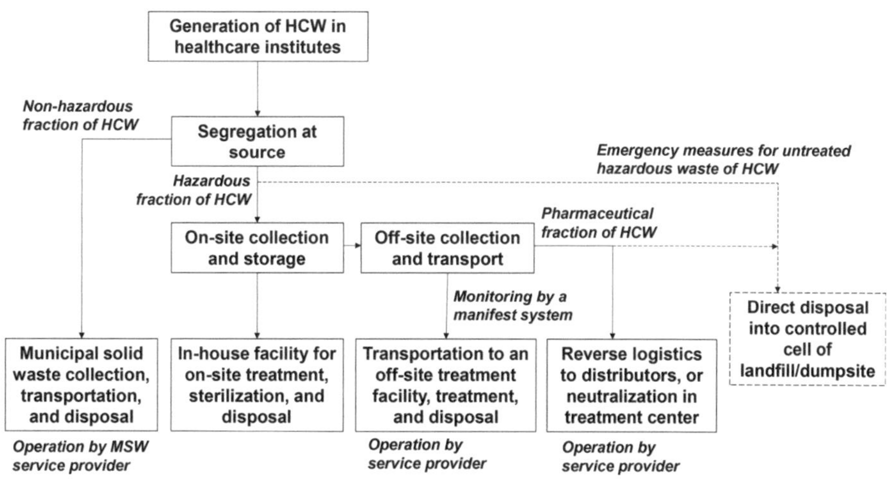

Figure 7.
Outline and options for HCW management in developing countries. Dashed parts are emergency measures under limited conditions, which is an example in Palestine.

3. Handling, storage, and transport

HCW workers have the greatest occupational risk (the Risk A in **Figure 3**), where the hazard is from direct contact with sharps and infectious waste. Sharps can cause puncture wounds, scratches, and scrapes, where infectious agents can penetrate the skin. The use of special containers for sharps is absolutely necessary. In the HCW handling process, there is also potential for exposure through inhalation of pathogen-containing aerosols or particulates [20].

The best way to minimize the risk of exposure is to ensure that the infectious waste is properly isolated. Some basic principles [20] are: packaging the infectious HCW properly; avoiding physical contact with the infectious HCW; using personal protective equipment (PPE); and handling the infectious HCW as little as possible.

Another factor to be considered is public health including informal waste pickers if the hazardous infectious waste is directly disposed without any treatment as an emergent measure under limited conditions. In that case, no one has easy access to discarded needles and syringes, so that waste sharps' containers need the following features: puncture resistance, impermeability, rigidity, tamper resistance, and proper marking [31].

4. Treatment technologies

The purpose of treatment is to change the biological character of infectious HCW to eliminate, or at least to significantly reduce, its potential for causing negative impacts. The three most common techniques used to treat infectious HCW are incineration (various types are available), steam sterilization (autoclaving), and microwaving (**Table 3**). Other currently available techniques include irradiation, chemical disinfection, and so on.

5. Disposal

When infectious HCW waste has been properly treated, the waste is no longer infectious. The treated HCW can be handled in the same way as normal municipal waste. However there are two exceptional cases [18]: for sharps and pathological wastes, additional processing before disposal is necessary; if other hazardous substances such as pharmaceutical, chemical, or radioactive waste are contained, there must be additional treatment before disposal.

In the case of sharps waste treated by steam sterilization, intact sharps are possibly sent to a landfill where workers are at risk for injury; therefore, they should be shredded or destroyed when be treated. Pathological waste treated by steam sterilization also requires additional processing since body parts or organs can be recognizable, which should not be directly disposed in the landfill.

In developing countries, sometimes no suitable HCW treatment facility is available and the only option is direct landfilling. In such a case, it is necessary to avoid using an open dumpsite and dispose a landfill having a shielded structure. It is necessary to immediately cover the soil at the time of waste disposal to prevent the dissipation of HCW and protect the environment.

6. Staff training

HCW management training is an effective intervention for preventing infections and improving the occupational safety of the HCWs through building awareness,

Treatment technology	Waste type	Operation & maintenance	Residues after the treatment	Environmental consideration
Incineration	• Applicable for almost all infectious HCW	• Relatively complex operation • It is necessary to segregate the highly flammable waste to avoid an explosion accident. • Requires skilled operators • Additional fuel is required for complete combustion	• Large volume reduction (approx. 90%)	• Smoke and gas emission with toxic substances are concerned when operated improperly • Relatively low to moderate cost
Autoclave (Steam sterilization)	• Applicable for most infectious HCW • Not applicable to pharmaceuticals	• Relatively simple operation • No additional fuel is required	• No volume reduction	• No emission when operated properly • Relatively moderate cost
Microwave	• Applicable for most infectious HCW • Not applicable for pharmaceuticals	• Moderately simple operation • Requires skilled operators • No additional fuel is required.	• Some volume reduction (max. 30%)	• No emission when operated properly • Relatively high cost
Direct Landfilling	• Applicable for almost all infectious HCW	• Simple operation • Landfill having shielded structure and daily covering soil are required.	• No volume reduction	• Possible migration of pollutants into the environment • Not suitable for long-term use

Table 3.
Comparison of most common HCW treatment technologies and landfilling [14, 20, 32–35].

changing attitudes and practices [36]. Training has two functions. One is to get a good understanding of the HCW management system so that the actual work can be implemented smoothly and without fail. The other is the meaning of risk communication. In other words, it is necessary for each worker to understand various risks involved in executing HCW management work and to give due consideration to safety. Training programs are planned for each job type, and manuals and posters are created as needed so that everyone can understand them. The staff training should be done continuously every year. Donor agencies are required to plan a training of trainer (ToT) for realizing a sustainable training program.

7. Monitoring

The monitoring system needs to be able to monitor the sorting status at the source, the amount and composition of the generation, the amount of treatment, and the amount of final disposal. In particular, it is necessary to track whether all infectious HCW has been properly collected and processed from the source.

5. Diagnosis of HCW management

In developing countries, depending on the conditions, the appearance of the problem in HCW management will differ, and some of the 14 problems mentioned in **Figure 6** and **Table 2**, combine to cause significant difficulties as a whole. Therefore, it is necessary to grasp the current situation of HCW, clarify the problems, and analyze the problems specifically.

A flowchart for diagnosing HCW management is shown in **Figure 8**. By answering 10 basic questions in the flowchart, required issues and challenges for capacity development in the HCW management are derived as follows:

1. Question 1 is 'Have the legal system and authority regarding HCW management been established?' A legal system on HCW must be in place that include definitions, treatment responsibilities, regulations, standards, and guidelines for HCW and its management. In addition, it is necessary to clarify the regulatory authority at the government level and the implementing body for HCW management at the national and local levels in accordance with the legal system. If the answer to this Question 1 is negative, then the challenge for capacity development is to establish a legal system, standards, and guidelines for HCW management and to determine regulatory agency at the government level.

2. Question 2 is 'Present state of HCW has been studied?' It is not possible to make an effective HCW management plan without an accurate understanding of the current state of HCW, in particular, the generation sources, generation amount, its composition, treatment, and disposal. If the answer to this Question 2 is negative, then the challenge for capacity development is to clarify the current waste stream of HCW if any, including the inventory of healthcare institutes, waste generation amount & composition, and treatment & disposal methods, through conducting a survey.

3. Question 3 is 'In-house HCW management system has been established?' In this question, the current state of HCW management at the individual healthcare

institute level is assessed based on the law and regulation, which includes the availability of HCW plan and also its practice including segregation of HCW at source, collection, storing, and treatment. If the answer to this Question 3 is negative, then the challenge for capacity development is to formulate an HCW management plan at each level, and establish an HCW management system and strengthen the capacity of the implementing organization. This question corresponds to Problem No.2.

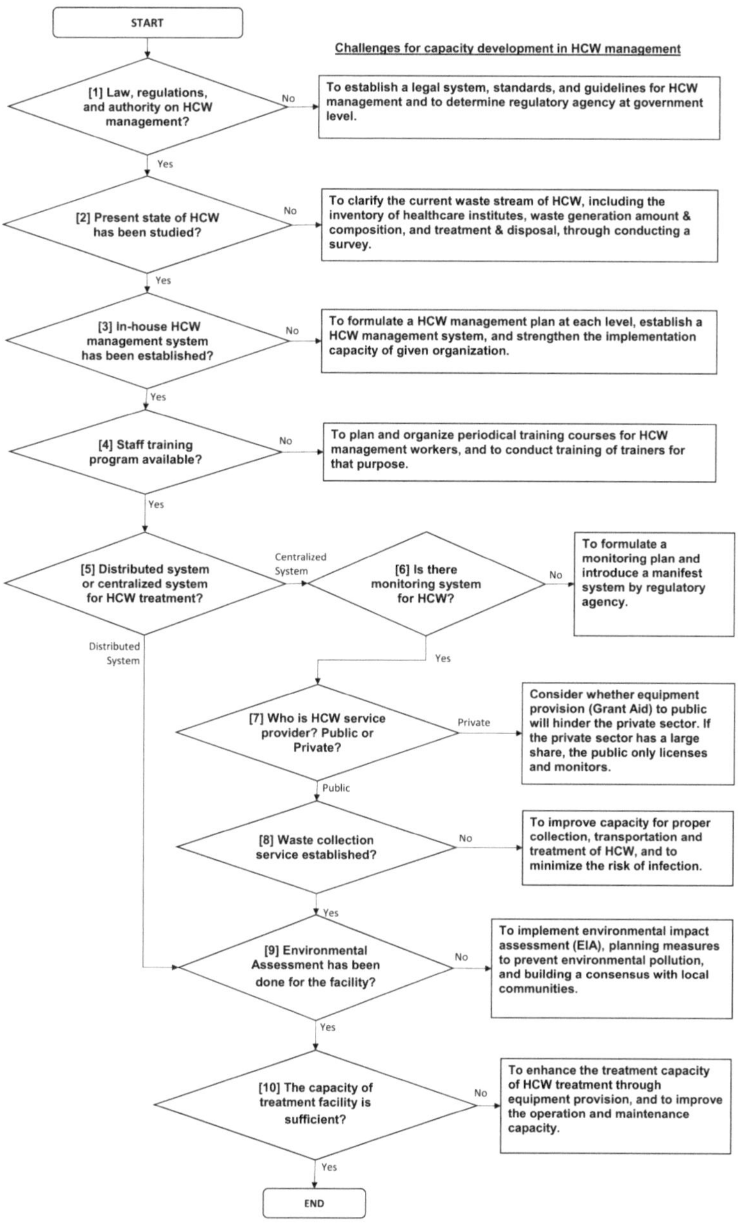

Figure 8.
A flowchart for diagnosing an HCW management system and for identifying challenges in capacity development, which are the targets of technical assistance.

4. Question 4 is 'Staff training program available?' Training and capacity building of staff responsible for the HCW management work is essential for effective HCW management. It is necessary that some kind of training function is set at each level of national, local, and healthcare institutes. If the answer to this Question 4 is negative, then the challenge for capacity development is to plan and organize periodical training courses for HCW management workers, and to conduct training of trainers for that purpose.

5. Question 5 is 'Distributed system or centralized system for HCW treatment?' It is a two-choice question. As mentioned above, the infectious HCW treatment system is either a distributed type that can be used in each healthcare facility or infectious HCW is separately discharged from each healthcare facility and treated at an external treatment facility, a centralized type. It is necessary to clarify which system the target country, area, or individual facility, is adopting or intending to adopt.

6. Question 6 is 'Is there monitoring system for HCW?' This question is for the case of centralized system. The monitoring system referred to here is to monitor whether infectious HCW is properly segregated, collected, and treated in the off-site facility. If the answer to this Question 6 is negative, then the challenge for capacity development is to formulate a monitoring plan and introduce a manifest system, which is designed to track infectious HCW from the time it leaves the generation source until it reaches the off-site facility by regulatory/monitoring agency.

7. Question 7 is 'Who is HCW service provider? Public or Private?' It is a two-choice question. Since infectious HCW treatment requires specialized technology, private specialized companies are often entrusted with treatment by healthcare institutions and perform the treatment as an off-site centralized system. On the other hand, public institutions also sometimes carry out the treatment in an off-site centralized system, which differs depending on each country and area. This question is important in particular for aid agencies because public intervention should not disrupt existing systems based on private enterprises' businesses, if private enterprises have a large share of the outsourced processing. For example, in an area where private services are predominant, if support such as the establishment of centralized treatment facilities is provided to public institutions as a grant aid project, it will clearly hinder competitiveness and create conflicts between public and private. Therefore, if the answer to the Question 7 is the 'private company', consider whether equipment provision (grant aid/donation) to the public will hinder the private sector. If the private sector services are existing, it is necessary to consider that the public conducts only licensing and monitoring to these private activities.

8. Question 8 is 'Waste collection service is well established?' This question is for a public SWM agency that employs a centralized system for collecting and treating infectious HCW. In such cases, it is important to consider whether public agencies are properly conducting SWM services. This is because the centralized system has a similar management structure to that of SWM; waste collection, transportation, treatment, and disposal. If the answer to this Question 8 is negative, then the challenge for capacity development is to improve implementing capacity of public agencies in SWM as well as HCW management.

9. Question 9 is 'Environmental Assessment has been done for the facility?' When installing an infectious HCW treatment facility, it is necessary to evaluate various environmental impacts such as gas emissions, wastewater, and solid waste in the treatment process. In addition, when installing a large centralized treatment facility, it is necessary to reach a consensus about the siting with surrounding local communities. If the answer to this Question 9 is negative, then the challenge for capacity development is to implement environmental impact assessment (EIA), planning measures to prevent environmental pollution, and building a consensus with local communities.

10. Question 10 is 'The capacity of treatment facility is sufficient?' Often, donor agencies ask this question first and precede the provision of facility and equipment, but as shown in the flowchart, the system only works with various technical soft components. If these soft components meet the requirements, government, or donor agencies can equip the facility and equipment, otherwise, sustainability cannot be ensured. To enhance the treatment capacity of HCW, it is also necessary to improve the operation and maintenance capacity.

6. Development of HCW management system in Palestine: a case study

Over the past decades of experience in Palestine is a typical example of inadequate HCW management [35], where one of the major threats came from that much of hazardous infectious HCW had mixed with MSW [37] and flowed into dumpsites without any treatment and safety measure [38, 39].

In 2010, the Palestinian National Authority (PNA) compiled a report on the development of a National Master Plan for Hazardous Waste Management in the West Bank and Gaza [40]. According to the report, only one-third of the healthcare facilities used special bags for HCW collection, whereas all other facilities consequently collected all types of HCW together with MSW, except for sharps that were being collected in special boxes. The report also stated that 80% of healthcare facilities in Gaza had no way to securely store HCW generated. MSW was generally collected by local government units (LGUs) without any discrimination between HCW and MSW, and eventually, all types of solid waste were mixed and disposed.

In 2012, PNA enacted Palestinian Authority Cabinet Decision No. (10) of 2012, "Medical Waste Management System, and its Uses". The bylaw allowed both distributed (on-site) and centralized (off-site) systems of HCW treatment. As a practice based on the bylaw, the first systematic HCW collection & treatment service has been started in the southern area of West Bank (Hebron and Bethlehem governorates), where an HCW treatment (microwave) facility was equipped under the support of the EU. It is a typical "centralized system" and its operation and maintenance (O/M), as well as HCW collection, transportation, and disposal, were conducted by the Hebron-Bethlehem Higher Joint Service Council for Solid Waste Management (H-B JSC). However, in the remaining middle and northern part of West Bank and over the Gaza Strip, most of the HCW generated was still mixed and disposed without any treatment. The required local HCW management plan was little formulated at each healthcare institution and authority.

Under the circumstances described above, international technical cooperation projects for the capacity development of Palestinian authorities on HCW management were organized in the Gaza Strip from 2015 [22, 39, 41, 42], those are composed of five components: (1) assessment study to grasp the current state of HCW; (2) formulation of a strategy and preparation of HCW management plan; (3) capacity building activities such as seminars, workshops, and staff training courses; (4) Pilot projects on on-site and off-site HCW management to verify the effectiveness, efficiency, and feasibility of the HCW management plan; and (5) Provision of equipment by international donors.

As of 2015, there were 2245 inpatient beds in public hospitals and 619 inpatient beds in private/NGO hospitals in the Gaza Strip. The proper on-site segregation of the infectious and sharps showed that 2.4–0.7 kg/day of HCW is generated from hospitals and clinics. The generation rate from outpatients accounts for a rate of 11.0 g–9.5 g per outpatient [41]. The estimated total HCW generation amount was around 7199 kg/day, and the estimated amount of infectious HCW was calculated around 1071 kg/day.

Healthcare institutions in the Gaza Strip are responsible for on-site management, where MOH conducts monitoring, supervision, and enforcing bylaw on all healthcare institutions for their compliance with appropriate on-site management; three categories of segregation at sources (sharps, infectious, and noninfectious), controlled storage, and separated discharge of HCW.

Regarding the responsibility for off-site HCW management in the centralized system, collection, transport, treatment by autoclave/microwave, and final disposal, the collection service have been managed by the Joint Service Council of Khan Yunisi, Rafah, and Middle Gaza (JSC-KRM) since 2017, and later JSC of North Gaza and Gaza (JSC-GNG) has started the service since 2020. An HCW management system was established over the Gaza Strip as the twin centralized systems using autoclave/microwave facilities for the infectious waste treatment (**Figure 9**). From 2017 to 2019, intensive training courses have been held for the workers in healthcare institutes and service providers (JSCs).

The costs of HCW management services are borne by the HCW generator based on the polluter-pay principle (PPP), where each healthcare institute pays a reasonable fee to JSCs. In the pilot project in the southern Gaza Strip from 2018 to 2019, the real costs for waste collection, transportation, treatment, and final disposal operations

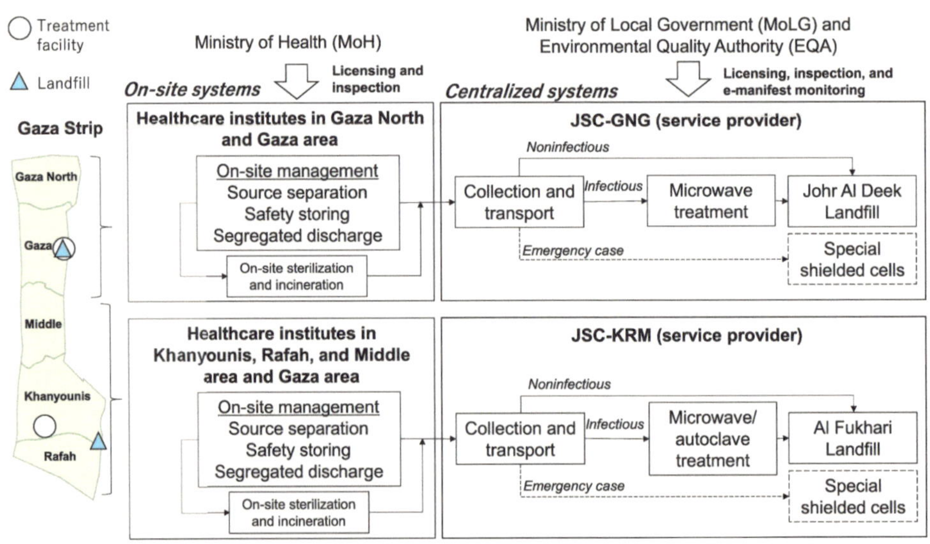

Figure 9.
Centralized HCW management systems, service providers, and supervising authorities in Gaza Strip.

were measured and aggregated, and the cost per unit weight of infectious HCW was determined. In this way, the HCW management systems in Gaza have basically been established.

HCW management is still not fully established in the West Bank area of Palestine. An international cooperation project supported by Japan (JICA) is currently conducting assessment surveys and planning, and in 2022, three centralized treatment facilities will be installed, staff training courses will be organized, and HCW management services will be started to the entire area.

More than 10-year process corresponds well with the 10-step diagnostic process described earlier (**Figure 8**), demonstrating that Palestine has gradually improved its capacity for HCW management.

In Gaza Strip, the 1st step was the enacting bylaw on HCW management in 2012, and then the 2nd step of the survey on the present state of HCW was conducted from 2010 (Master Plan) to 2016 (Gaza Local Plan). The 3rd step of the local HCW management plan and 4th step of the staff training program was conducted from 2017 to 2019. Under given conditions, public (JSCs) operating centralized treatment system was introduced as 5th and 6th steps in 2018, and autoclave and microwave facilities were equipped from 2018 (autoclave) to 2021 (microwave) as the 9th to 10th steps.

On the West Bank, it is still in the 2nd and 3rd steps, but in 2022, the treatment facility will be equipped, training will be conducted, HCW management services will be started, and 10 steps are expected to be achieved.

7. Conclusions

Proper management of HCW in developing countries is an urgent issue. It is important not only for public health and environmental protection in developing countries, but also for controlling infections throughout the world, in the time of pandemics.

According to statistical cross-country analysis, the amount of HCW generation shows a moderate positive power correlation with economic growth (GDP/capita). It also shows a positive power correlation with the amount of MSW generated. This indicates that economic growth will lead to an increase in MSW as well as a rapid increase in the amount of HCW generated. On the other hand, in countries with a high level of service coverage of MSW collection of 83% or more, a sudden increase in the amount of HCW are observed. This is considered to indicate that the recognition of HCW generation and the necessity of its proper management are formed in the public administration of HCW issues when the MSW management service reaches a certain stage.

It is necessary to establish an effective HCW management system and strengthen its implementation capacity, especially in developing countries. When conducting international technical assistance for them to support the establishment of the HCW management system and capacity development, it is required to set the targets for technical assistance through conducting an assessment survey, analyzing the problems, evaluating risks, supporting to formulate HCW management plans, and provision of equipment.

A flow chart to the 10-step diagnosis of HCW management is proposed to identify issues and challenges of HCW management in developing countries. The results can be used for setting the targets of technical assistance and cooperation for enhancing HCW management. More than 10-year process of technical assistance and cooperation program in Gaza Strip, Palestine corresponds well with the 10-step diagnostic process, demonstrating that Palestine has gradually improved its capacity for HCW management.

Acknowledgements

The author expresses special thanks to Hideaki Matsuoka and Chie Shimodaira of JICA Global Environment Department for their valuable discussions on the diagnosis of HCW management in developing countries. The author also thanks Suleiman Abu Mufarreh of Ministry of Local Government, Palestine, Reem Abukmeil of former MoLG-JICA project coordinator, Majdi Dher of Ministry of Health, Ali Bahoum of JSC-KRM, and Yuko Mitsui, Yuko Santo, Mariko Chiba, Saher Jaber Younis of JICA Palestine for their supports, during the implementation of technical assistance program on HCW management in Gaza Strip, Palestine. The view expressed in this chapter does not necessarily reflect the official positions of JICA or Palestinian Authority.

Author details

Mitsuo Yoshida
Global Environment Department, Japan International Cooperation Agency (JICA), Nibancho, Chiyoda-ku, Tokyo, Japan

*Address all correspondence to: yoshida.mitsuo@jica.go.jp

References

[1] Chartier Y, Emmanuel J, Pieper U, Prüss A, Rushbrook P, Stringer R, et al. Safe Management of Wastes from Health-Care Activities. 2nd ed. Geneva: World Health Organization (WHO); 2014. p. 308. ISBN 978 92 4 154856 4

[2] US-EPA. What is a Hazardous Waste? Learn the Basics of Hazardous Waste. United States Environmental Protection Agency [Internet]. Available from: https://www.epa.gov/hw/learn-basics-hazardous-waste [Accessed: September 21, 2021]

[3] Jang YC. Infectious/medical/hospital waste: General characteristics. In: Nriagu J, editor-in-Chief. Encyclopedia of Environmental Health. 2nd ed. Amsterdam: Elsevier; 2011. pp. 661-665. DOI: 10.1016/B978-0-444-63951-6.00508-8

[4] Win EM, Saw YM, Oo KL, Than TM, Cho SM, Kariya T, et al. Healthcare waste management at primary health centres in Mon State, Myanmar: The comparisons between hospital and non-hospital type primary health centres. Nagoya Journal of Medical Science. 2019;**81**(1):81-91. DOI: 10.18999/nagjms.81.1.81

[5] Amsalu A, Worku M, Tadesse E, Shimelis T. The exposure rate to hepatitis B and C viruses among medical waste handlers in three government hospitals, southern Ethiopia. Epidemiology and Community Health. 2016;38. Article ID: e2016001. 8 pages. DOI: 10.4178/epih/e2016001

[6] Singh N, Tang Y, Zhang Z, Zheng C. COVID-19 waste management: Effective and successful measures in Wuhan, China. Resources, Conservation & Recycling. 2020;**163**:105071. DOI: 10.1016/j.resconrec.2020.105071

[7] Thakur V. Framework for PESTEL dimensions of sustainable healthcare waste management: Learnings from COVID-19 outbreak. Journal of Cleaner Production. 2021;**287**:125562

[8] Kampf G, Todt D, Pfaender S, Steinmann E. Persistence of coronaviruses on inanimate surfaces and their inactivation with biocidal agents. Journal of Hospital Infection. 2020;**104**:246-251. DOI: 10.1016/j.jhin.2020.01.022

[9] United Nations. World Economic Situation and Prospects 2021. New York: Department of Economic and Social Affairs, United Nations; 2021. 172 p. ISBN: 978-92-1-109182-3

[10] WHO/UNICEF. Water, Sanitation and Hygiene in Health Care Facilities Status in Low- and Middle-income Countries and Way Forward. Geneva: World Health Organization; 2015. ISBN: 978 92 4 150847 6

[11] Secretariat of the Basel Convention (SBC) and WHO. Preparation of National Health-Care Waste Management Plans in Sub-Saharan Countries, Guidance Manual. Amsterdam: World Health Organization (WHO); 2005. p. 88. ISBN: 92 4 154662 X

[12] Sarsour A, Ayoub A, Lubbad I, Omran A, Shahrour I. Assessment of medical waste management within selected hospitals in Gaza strip Palestine: A pilot study. International Journal of Scientific Research in Environmental Science and Toxicology. 2014;**2**(5):164-173. DOI: 10.12983/ijsres-2014-p0164-0173

[13] Nzediegwu C, Chang SX. Improper solid waste management increases potential for COVID-19

spread in developing countries. Resources, Conservation & Recycling. 2020;**161**:104947

[14] Diaz LF, Savage GM, Eggerth LL. Alternatives for the treatment and disposal of healthcare wastes in developing countries. Waste Management. 2005;**25**:626-637

[15] Abah SO, Ohimain EI. Healthcare waste management in Nigeria: A case study. Journal of Public Health and Epidemiology. 2011;**3**(3):99-110

[16] Kuchibanda K, Mayo AW. Public health risks from mismanagement of healthcare wastes in Shinyanga municipality health facilities, Tanzania. The Scientific World Journal. 2015;**2015**:11. Article ID: 981756. DOI: 10.1155/2015/981756

[17] Nwachukwu NC, Orji FA, Ugbogu OC. Health care waste management—Public health benefits, and the need for effective environmental regulatory surveillance in Federal Republic of Nigeria. In: Current Topics in Public Health, edited by Alfonso Rodriguez-Morales. London, UK: IntechOpen; 2013. DOI: 10.5772/53196

[18] Haylamicheal ID, Desalegne SA. A review of legal framework applicable for the management of healthcare waste and current management practices in Ethiopia. Waste Management & Research. 2011;**30**(6):607-618

[19] Kaza S, Yao L, Bhada-Tata P, Van Woerden F. What a Waste 2.0: A Global Snapshot of Solid Waste Management to 2050. Urban Development Series. Washington, DC: The World Bank; 2018. p. 272. DOI: 10.1596/978-1-4648-1329-0

[20] Reinhardt PA, Gordon JG. Infectious and Medical Waste Management. New York: Lewis Publ.; 1991. p. 280

[21] Salkin IF. Review of Health Impacts from Microbiological Hazards in Health-Care Wastes. WHO, Geneva: Department of Blood Safety and Clinical Technology and Department of Protection of the Human Environment; 2004

[22] MoLG-JICA. Final Report on the Project for Technical Assistance in Solid Waste management in Palestine (Phase 2). Tokyo: Japan International Cooperation Agency (JICA); 2019

[23] Centers for Disease Control and Prevention (CDC). Guidelines for Environmental Infection Control in Health-Care Facilities [Internet]. 2003. Available from: https://www.cdc.gov/infectioncontrol/guidelines/environmental/background/medical-waste.html#i6 [Accessed: December 01, 2021]

[24] Collins CH, Kennedy DA. The microbiological hazards of municipal and clinical wastes. Journal of Applied Microbiology. 1992;**73**:1-6

[25] Dragon DC, Rennie RP. The ecology of anthrax spores: Tough but not invincible. The Canadian Veterinary Journal. 1995;**36**(5):295-301

[26] WHO. Overview of Technologies for the Treatment of Infectious and Sharp Waste from Health Care Facilities. Geneva: WHO; 2019. 42 p

[27] WHO. Guidelines for the Safe Disposal of Unwanted Pharmaceuticals in and after Emergencies Interagency Guidelines. WHO/EDM/PAR/99.2; 1999. 31 p

[28] Townend WK, Cheeseman C, Edgar J, Tudor T. Factors driving the development of healthcare waste management in the United Kingdom over the past 60 years. Waste Management & Research. 2009;**27**:362-373

[29] Global Healthcare Waste Project. MODULE 3: International and National HCWM Laws—Legislative, Regulatory and Policy Aspect [Internet]. Available from: https://cdn.who.int/media/docs/default-source/wash-documents/wash-in-hcf/training-modules-in-health-care-waste-management/module-3---international-and-national-hcwm-laws.pdf

[30] United Nations Conference on Environment and Development in Rio de Janeiro. The Principle 15 of the Rio Declaration on Environment and Development. New York: United Nations; 1992

[31] OSHA (US Occupational Safety and Health Administration). OSHA's Guidelines for Bloodborne Pathogens, Waste, Sharps, and Catheter [Internet]. Available from: https://storage.googleapis.com/edcompass/material/OSHA_Bloodborne_Pathogens_Waste_Sharps_and_Catheters.pdf [Accessed: December 01, 2021]

[32] WHO. 8. Treatment and Disposal Technologies for Health-care Waste [Internet]. Available from: https://www.who.int/water_sanitation_health/medicalwaste/077to112.pdf [Accessed: December 01, 2021]

[33] WHO. 9. Application of Treatment and Disposal Methods to Health-care Waste Categories [Internet]. Available from: https://www.who.int/water_sanitation_health/medicalwaste/113to129.pdf [Accessed: December 01, 2021]

[34] Cross FL, Hesketh HE, Rykowski PK. Infectious Waste Management. Lanchester: Technomic Publ; 1990. 150 p

[35] Al Khatib I, Al-Qaroot YS, Shtayeh MS. Management of healthcare waste in circumstances of limited resources: A case study in the hospitals of Nablus city, Palestine. Waste Management & Research. 2009;**27**:305-312. DOI: 10.1177/0734242X08094124

[36] Ara L, Vashkar SMK, Mowla M, Hossain MD, Mondal UK. Preventing infections and improving occupational safety among the healthcare workers through intensive healthcare waste management training in a developing country. American Journal of Infection Control. 2016;**44**:S14

[37] UNDP. Feasibility study and detailed design for solid waste management in the Gaza Strip. Deliverable of the project feasibility study and detailed design for short-term and long-term solid waste management in the Gaza Strip [Internet]. 2012. Available from: http://www.mdlf.org.ps/Files/solid-waste-gaza/Al%20Fukhari%20Landfill%20Operational%20Manual.pdf [Accessed: April 02, 2020]

[38] WHO Specialized Health Mission to the Gaza Strip. As Mandated by the Executive Board of WHO in its Resolution EB124.R4 [Internet]. 2009. Available from: https://www.who.int/hac/crises/international/wbgs/gaza_specialized_mission_extendedrep_21may09.pdf [Accessed: April 07, 2020]

[39] Caniato M, Tudor TL, Vaccari M. Understanding the perceptions, roles and interactions of stakeholder networks managing health-care waste: A case study of the Gaza Strip. Waste Management. 2015;**35**(January):255-264. DOI: 10.1016/j.wasman.2014.09.018

[40] El-Hamouz A. The Development of a National Master Plan for Hazardous Waste Management for the Palestinian National Authority (PNA). Ramallah: Palestinian National

Authority; [Internet]. 2010. Available from: https://environment.pna.ps/ar/files/Part_one_Final_Report_on_The_Development_of_a_National_Master_Plan_for_Hazardous_Waste_Management_for_the_Palestinian_National_Authority_en.pdf [Accessed: April 02, 2020]

[41] Abukmeil R, Barhoum A, Dher M, Yoshida M. Assessment of health-care waste generation and its management strategy in the Gaza Strip, Palestine. Journal of Material Cycles and Waste Management. 2021. DOI: 10.1007/s10163-021-01306-4

[42] Caniato M, Tudor TL, Vaccari M. Assessment of health-care waste management in a humanitarian crisis: A case study of the Gaza Strip. Waste Management. 2016;**58**:386-396. DOI: 10.1016/j.wasman.2016.09.017